AF707808

Method of Moments for 2D Scattering Problems

FOCUS SERIES

Series Editor Joseph Saillard

Method of Moments for 2D Scattering Problems

Basic Concepts and Applications

Christophe Bourlier
Nicolas Pinel
Gildas Kubické

iSTE

WILEY

First published 2013 in Great Britain and the United States by ISTE Ltd and John Wiley & Sons, Inc.

Apart from any fair dealing for the purposes of research or private study, or criticism or review, as permitted under the Copyright, Designs and Patents Act 1988, this publication may only be reproduced, stored or transmitted, in any form or by any means, with the prior permission in writing of the publishers, or in the case of reprographic reproduction in accordance with the terms and licenses issued by the CLA. Enquiries concerning reproduction outside these terms should be sent to the publishers at the undermentioned address:

ISTE Ltd
27-37 St George's Road
London SW19 4EU
UK

www.iste.co.uk

John Wiley & Sons, Inc.
111 River Street
Hoboken, NJ 07030
USA

www.wiley.com

© ISTE Ltd 2013
The rights of Christophe Bourlier, Nicolas Pinel and Gildas Kubické to be identified as the authors of this work have been asserted by them in accordance with the Copyright, Designs and Patents Act 1988.

Library of Congress Control Number: 2013941777

British Library Cataloguing-in-Publication Data
A CIP record for this book is available from the British Library
ISSN: 2051-2481 (Print)
ISSN: 2051-249X (Online)
ISBN: 978-1-84821-472-9

MIX
Paper from responsible sources
FSC® C013604

Printed and bound in Great Britain by CPI Group (UK) Ltd., Croydon, Surrey CR0 4YY

Contents

Preface

Electromagnetic wave scattering from randomly rough surfaces in the presence of scatterers is an active, interdisciplinary area of research with countless practical applications in fields such as optics, acoustics, geoscience and remote sensing. In the last four decades, considerable theoretical progress has been made in elucidating and understanding the scattering processes involved in such problems. Numerical simulations allow us to solve the Maxwell equations exactly, without the limitations of asymptotic approaches whose regimes of validity are often difficult to assess. The purpose of this book is to present both asymptotic approaches, such as the Kirchhoff approximation, and numerical methods, such as the method of moments (MoM), in order to solve scattering from rough surfaces.

Excellent textbooks on this subject are available and this book focuses on some scattering problems such as the scattering from a rough surface, a rough layer, a coated cylinder and an object near a rough surface. Although the scattering problem is assumed to be two-dimensional (invariant with respect to a direction), the problem is of practical interest because large problems can easily be solved, unlike a direct three-dimensional scattering problem, for which the equations are more complicated (because they are vectorial). Indeed, due to computing time and memory space requirements, the size of the problem to be solved is reduced. Nevertheless, advanced numerical methods can handle large problems, but the complexity of programming significantly increases.

This book is intended both for graduate students who wish to learn about scattering by rough surfaces and engineers or researchers who have to solve such problems. Adding a scatterer near a rough surface, from the MoM, the problem size increases significantly and in order to solve this problem using a standard personal computer, in Chapters 3 and 4, a versatile method, which has been developed in the last decade, is presented in detail.

The increasingly important role of numerical simulations in solving electromagnetic wave scattering problems has motivated us to provide the readers with computer codes on topics relevant to the book. These computer codes are written in the MatLab programming language. They are provided for two main purposes. The first purpose is to provide the readers a hands-on training for performing numerical experiments, through which the concepts of the book can be better communicated. The second purpose is to give new researchers a set of basic tools with which they could quickly build on their own projects.

To have the MatLab programs, please send an email to Dr. C. Bourlier at christophe.bourlier@univ-nantes.fr by providing a receipt of the purchase of this book.

My thanks go to several people who made this book possible. I am grateful to Professor Joseph Saillard (retired), for suggesting writing this book, and both Professors Saillard and Serge Toutain (retired) for giving me the means to develop this research. I would like to acknowledge Drs N. Déchamps and G. Kubické, the PhD students whom I co-supervised and who developed the PILE and E-PILE methods thoroughly presented in this book. I would also like to thank the National Center for Scientific Research by whom I am employed, and the DGA (*Direction Générale de l'Armement*) for their financial support.

Christophe BOURLIER
June 2013

Introduction

In this book, the method of moments (MoM) is addressed to compute the field scattered by scatterers such as canonical objects (cylinder or plate) or a randomly rough surface, and also by an object above or below a random rough surface. Because the problem is considered two-dimensional (2D), the integral equations (IEs) are scalar and only the transverse electric (TE) and transverse magnetic (TM) polarizations are considered (no cross polarizations occur). Chapter 1 analyzes how the MoM with the point-matching method and pulse basic functions is applied to convert the IEs into a linear system. In addition, Chapter 1 presents the statistical parameters necessary to generate Gaussian random rough surfaces. Chapter 2 compares the MoM with the exact solution of the field scattered by a circular cylinder in free space, and with the physical optics (PO) approximation for the scattering from a plate in free space. Chapter 3 presents numerical results, obtained from the MoM combined with the efficient E-PILE method, of the scattering from two illuminated scatterers and shows how the E-PILE algorithm can be hybridized with asymptotic or rigorous methods valid for the scattering from a single scatterer (alone). Chapter 4 presents the same results as those in Chapter 3 but for an object above a random rough surface or for a coated (circular or elliptical) cylinder. In the last two chapters, the coupling between the two scatterers is also studied in detail by inverting the impedance matrix by blocks.

1

Integral Equations for a Single Scatterer: Method of Moments and Rough Surfaces

1.1. Introduction

In this chapter, the integral equations (IEs) are addressed to derive the field scattered by a single scatterer in free space. They are obtained by introducing the Green function concept and by applying the boundary conditions onto the scatterer. In addition, the IEs are converted into a linear system by using the method of moments (MoM) with the point-matching method. The impedance matrix is then expressed for any shape of the object. The special case of a perfectly conducting (PC) object is also discussed. This chapter also presents the necessary statistical parameters to generate a random rough surface.

In all chapters, the media are considered as homogeneous, linear and isotropic. In addition, they are considered as non-magnetic, which means that the magnetic permeability is $\mu_0 = 4\pi \times 10^{-7}$ H/m. In addition, the medium Ω_0 (the subscript 0 is used for variables defined in vacuum) is considered as vacuum and the time convention $e^{-j\omega t}$ is used. Then, the derivative over the time t is $\partial/\partial t \rightarrow -j\omega$. For any media without sources, two Maxwell equations [KON 05, TSA 00] are simplified as:

$$\begin{cases} \textbf{curl}\, \boldsymbol{H} = \nabla \wedge \boldsymbol{H} = -j\omega\epsilon \boldsymbol{E} \\ \textbf{curl}\, \boldsymbol{E} = \nabla \wedge \boldsymbol{E} = j\omega\mu_0 \boldsymbol{H} \end{cases}, \qquad [1.1]$$

where $\boldsymbol{H}$ is the magnetic field and $\boldsymbol{E}$ the electric field. In addition, ω is the pulsation (rad/s) and $\epsilon = \epsilon_0\epsilon_r$ is the permittivity, in which $\epsilon_0 = 1/(36\pi \times 10^9)$

is the permittivity in vacuum and ϵ_r is the relative permittivity (which equals unity for vacuum). For a two-dimensional (2D) space of unitary vectors $(\hat{\boldsymbol{x}}, \hat{\boldsymbol{z}})$, the vectorial operator $\boldsymbol{\nabla}$ is defined in Cartesian coordinates as:

$$\boldsymbol{\nabla} = \frac{\partial}{\partial x}\hat{\boldsymbol{x}} + \frac{\partial}{\partial z}\hat{\boldsymbol{z}}. \quad [1.2]$$

1.2. Integral equations

1.2.1. *TE and TM polarizations and boundary conditions*

Let $\hat{\boldsymbol{n}}$ be the normal to the surface S pointing toward Ω_0 and lying in the plane $(\hat{\boldsymbol{x}}, \hat{\boldsymbol{z}})$ (2D problem), and separating two media, Ω_0 (upper) and Ω_1 (lower), of dielectric permittivities ϵ_0 and ϵ_1 (see Figure 1.1), respectively.

For the transverse electric (TE) polarization (the electric field is normal to the incident plane $(\hat{\boldsymbol{x}}, \hat{\boldsymbol{z}})$), the electric field in the upper medium is defined as $\boldsymbol{E}_0 = \psi_0\hat{\boldsymbol{y}}$, where ψ_0 is a scalar number. In medium Ω_0, the use of equation [1.1] leads to:

$$\boldsymbol{H}_0 = \frac{1}{j\omega\mu_0}\boldsymbol{\nabla} \wedge \boldsymbol{E}_0 = -\frac{1}{j\omega\mu_0}\hat{\boldsymbol{y}} \wedge \boldsymbol{\nabla}\psi_0. \quad [1.3]$$

knowing that $\boldsymbol{A}_1 \wedge (\boldsymbol{A}_2 \wedge \boldsymbol{A}_3) = (\boldsymbol{A}_1 \cdot \boldsymbol{A}_3)\boldsymbol{A}_2 - (\boldsymbol{A}_1 \cdot \boldsymbol{A}_2)\boldsymbol{A}_3$, for any vectorial function $\boldsymbol{A}_i$, we have:

$$\begin{aligned}\hat{\boldsymbol{n}} \wedge \boldsymbol{H}_0 &= -\frac{1}{j\omega\mu_0}\hat{\boldsymbol{n}} \wedge (\hat{\boldsymbol{y}} \wedge \boldsymbol{\nabla}\psi_0) = -\frac{1}{j\omega\mu_0}[(\hat{\boldsymbol{n}} \cdot \boldsymbol{\nabla}\psi_0)\hat{\boldsymbol{y}} - (\hat{\boldsymbol{n}} \cdot \hat{\boldsymbol{y}})\boldsymbol{\nabla}\psi_0] \\ &= -\frac{\hat{\boldsymbol{y}}}{j\omega\mu_0}(\hat{\boldsymbol{n}} \cdot \boldsymbol{\nabla}\psi_0), \end{aligned} \quad [1.4]$$

where $\hat{\boldsymbol{n}} \cdot \hat{\boldsymbol{y}} = 0$ since the normal lies in the plane $(\hat{\boldsymbol{x}}, \hat{\boldsymbol{z}})$. For the lower medium Ω_1, the quantities $\boldsymbol{E}_1$, $\boldsymbol{H}_1$ and ψ_1 also satisfy equation [1.4], in which the subscript "1" is used for variables defined in Ω_1.

For a surface separating two dielectric media, the boundary conditions state that the electric and magnetic tangential fields are continuous. Since $\boldsymbol{E}_{0,1} = \psi_{0,1}\hat{\boldsymbol{y}}$, this leads from equation [1.4] to:

$$\begin{cases}\psi_0(\boldsymbol{r}) = \psi_1(\boldsymbol{r}) \\ \hat{\boldsymbol{n}} \cdot \boldsymbol{\nabla}\psi_0(\boldsymbol{r}) = \hat{\boldsymbol{n}} \cdot \boldsymbol{\nabla}\psi_1(\boldsymbol{r})\end{cases}, \quad \forall \boldsymbol{r} \in S. \quad [1.5]$$

For the transverse magnetic (TM) polarization (the magnetic field is normal to the incidence plane $(\hat{\boldsymbol{x}}, \hat{\boldsymbol{z}})$), the magnetic field in the upper medium is defined as $\boldsymbol{H}_0 = \psi_0\hat{\boldsymbol{y}}$. The use of equation [1.1] leads to:

$$\boldsymbol{E}_0 = -\frac{1}{j\omega\epsilon_0}\boldsymbol{\nabla}\wedge\boldsymbol{H}_0 = \frac{1}{j\omega\epsilon_0}\hat{\boldsymbol{y}}\wedge\boldsymbol{\nabla}\psi_0, \qquad [1.6]$$

and

$$\begin{aligned}\hat{\boldsymbol{n}}\wedge\boldsymbol{E}_0 &= -\frac{1}{j\omega\epsilon_0}(\hat{\boldsymbol{y}}\wedge\boldsymbol{\nabla}\psi_0)\wedge\hat{\boldsymbol{n}} = +\frac{1}{j\omega\epsilon_0}\hat{\boldsymbol{n}}\wedge(\hat{\boldsymbol{y}}\wedge\boldsymbol{\nabla}\psi_0)\\ &= -\frac{1}{j\omega\epsilon_0}\left[(\hat{\boldsymbol{n}}\cdot\boldsymbol{\nabla}\psi_0)\hat{\boldsymbol{y}} - (\hat{\boldsymbol{n}}\cdot\hat{\boldsymbol{y}})\boldsymbol{\nabla}\psi_0\right]\\ &= +\frac{\hat{\boldsymbol{y}}}{j\omega\epsilon_0}(\hat{\boldsymbol{n}}\cdot\boldsymbol{\nabla}\psi_0). \end{aligned} \qquad [1.7]$$

Moreover, for the lower medium Ω_1, $\boldsymbol{H}_1 = \psi_1\hat{\boldsymbol{y}}$ and $\hat{\boldsymbol{n}}\wedge\boldsymbol{E}_1 = +\hat{\boldsymbol{y}}(\hat{\boldsymbol{n}}\cdot\boldsymbol{\nabla}\psi_1)/(j\omega\epsilon_1)$. The boundary conditions state that the electric and magnetic tangential fields are continuous, leading to:

$$\begin{cases}\psi_0(\boldsymbol{r}) = \psi_1(\boldsymbol{r})\\ \hat{\boldsymbol{n}}\cdot\boldsymbol{\nabla}\psi_0(\boldsymbol{r}) = \frac{\epsilon_0}{\epsilon_1}\hat{\boldsymbol{n}}\cdot\boldsymbol{\nabla}\psi_1(\boldsymbol{r})\end{cases}, \quad \forall\boldsymbol{r}\in S. \qquad [1.8]$$

In conclusion, for the TE and TM polarization, equations [1.5] and [1.8] lead to:

$$\begin{cases}\psi_0(\boldsymbol{r}) = \psi_1(\boldsymbol{r})\\ \hat{\boldsymbol{n}}\cdot\boldsymbol{\nabla}\psi_0(\boldsymbol{r}) = \rho_{01}\hat{\boldsymbol{n}}\cdot\boldsymbol{\nabla}\psi_1(\boldsymbol{r})\end{cases}, \quad \forall\boldsymbol{r}\in S, \qquad [1.9]$$

where $\rho_{01} = 1$ for the TE polarization and $\rho_{01} = \epsilon_0/\epsilon_1$ for the TM polarization.

1.2.2. *Electric and magnetic currents for a 2D problem*

For a 3D problem, the electric $\boldsymbol{J}_0$ and magnetic $\boldsymbol{M}_0$ currents are defined on the surface as:

$$\boldsymbol{J}_0 = \hat{\boldsymbol{n}}\wedge\hat{\boldsymbol{H}}_0, \; \boldsymbol{M}_0 = -\hat{\boldsymbol{n}}\wedge\hat{\boldsymbol{E}}_0, \qquad [1.10]$$

where $\hat{\boldsymbol{n}}$ is the normal to the surface.

For the TE polarization, $\boldsymbol{E}_0 = \psi_0 \hat{\boldsymbol{y}}$ and from equation [1.4], we then have:

$$\begin{cases} \boldsymbol{J}_0 = -\frac{1}{j\omega\mu_0}\hat{\boldsymbol{n}} \wedge (\hat{\boldsymbol{y}} \wedge \boldsymbol{\nabla}\psi_0) = -\frac{1}{j\omega\mu_0}(\hat{\boldsymbol{n}} \cdot \boldsymbol{\nabla}\psi_0)\,\hat{\boldsymbol{y}} \\ \boldsymbol{M}_0 = -\hat{\boldsymbol{n}} \wedge (\psi_0 \hat{\boldsymbol{y}}) = \psi_0 (-\hat{\boldsymbol{n}} \wedge \hat{\boldsymbol{y}}) \end{cases}. \quad [1.11]$$

For the TM polarization, $\boldsymbol{H}_0 = \psi_0 \hat{\boldsymbol{y}}$ and from equation [1.7], we then have:

$$\begin{cases} \boldsymbol{J}_0 = \hat{\boldsymbol{n}} \wedge (\psi_0 \hat{\boldsymbol{y}}) = \psi_0 (\hat{\boldsymbol{n}} \wedge \hat{\boldsymbol{y}}) \\ \boldsymbol{M}_0 = -\frac{1}{j\omega\epsilon_0}\hat{\boldsymbol{n}} \wedge (\hat{\boldsymbol{y}} \wedge \boldsymbol{\nabla}\psi_0) = -\frac{1}{j\omega\epsilon_0}(\hat{\boldsymbol{n}} \cdot \boldsymbol{\nabla}\psi_0)\,\hat{\boldsymbol{y}} \end{cases}. \quad [1.12]$$

In conclusion, for the TE and TM polarizations, ψ_0 and the normal derivative $\hat{\boldsymbol{n}} \cdot \boldsymbol{\nabla}\psi_0 = \partial\psi_0/\partial n$ are related to the currents $\{\boldsymbol{M}_0, \boldsymbol{J}_0\}$ and $\{\boldsymbol{J}_0, \boldsymbol{M}_0\}$, respectively.

1.2.3. *Huygens' principle and extinction theorem*

In Figure 1.1, the upper medium Ω_0 stands for the domain bounded by the surface S and the contour $C_{0,\infty}$, whereas Ω_1 stands for the domain bounded by the surface S and the contour $C_{1,\infty}$. We recall that the normal to the surface $\hat{\boldsymbol{n}}$ pointed toward Ω_0.

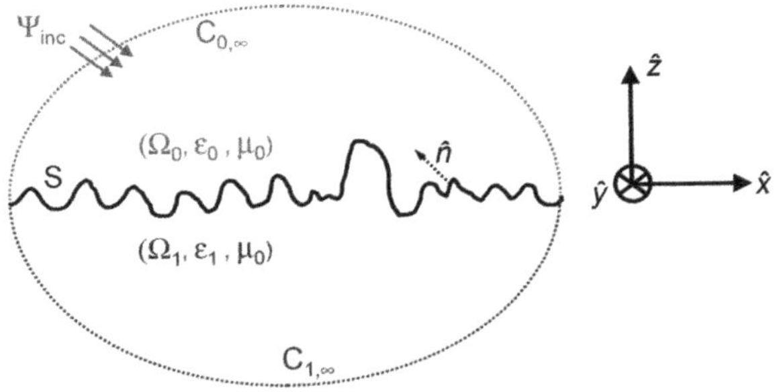

Figure 1.1. *The domain Ω_0 is bounded by the contour $C_{0,\infty}$ and the surface S whereas Ω_1 is bounded by the contour $C_{1,\infty}$ and the surface S*

In media Ω_0 and Ω_1 (without sources), the fields ψ_0 and ψ_1 satisfy the scalar Helmholtz equation:

$$\nabla^2 \psi_i(\boldsymbol{r}) + k_i^2 \psi_i(\boldsymbol{r}) = 0, \ \forall \boldsymbol{r} \in \Omega_i, \qquad [1.13]$$

where $k_i = \omega\sqrt{\epsilon_i \mu_0}$ is the wave number in medium Ω_i ($i = \{0, 1\}$) and ∇^2 is the scalar Laplacian. The scalar Green functions $g_0(\boldsymbol{r}, \boldsymbol{r}')$ and $g_1(\boldsymbol{r}, \boldsymbol{r}')$ defined in media Ω_0 and Ω_1, respectively, satisfy:

$$\nabla^2 g_i(\boldsymbol{r}, \boldsymbol{r}') + k_i^2 g_i(\boldsymbol{r}, \boldsymbol{r}') = -\delta(\boldsymbol{r} - \boldsymbol{r}'), \qquad [1.14]$$

where $\boldsymbol{r} = x\hat{\boldsymbol{x}} + z\hat{\boldsymbol{z}}$ are the Cartesian coordinates of the source point and $\boldsymbol{r}' = x'\hat{\boldsymbol{x}} + z'\hat{\boldsymbol{z}}$ are the Cartesian coordinates of the observation point. Physically, the Green function is the field radiated from a source point represented from the Dirac delta function δ. It then satisfies the Helmholtz equation, in which the right-hand side of equation [1.13] is $-\delta(\boldsymbol{r} - \boldsymbol{r}')$.

Applying the scalar Green theorem, we have:

$$\iint_\Omega (f_1 \nabla^2 f_2 - f_2 \nabla^2 f_1) dr = \oint_C (f_1 \boldsymbol{\nabla} f_2 - f_2 \boldsymbol{\nabla} f_1) \cdot \boldsymbol{dS}, \qquad [1.15]$$

where $\boldsymbol{dS} = \hat{\boldsymbol{n}} dS$, in which $\hat{\boldsymbol{n}}$ is the unitary vector normal to the closed oriented contour bounding medium Ω.

Let $f_1 = \psi_0$ and $f_2 = g_0$ in equation [1.15]. Then, from equations [1.13] and [1.14], for $\boldsymbol{r} \in \Omega_0$ we have:

$$\begin{aligned}
&\iint_{\Omega_0} \left[\psi_0(\boldsymbol{r}) \nabla^2 g_0(\boldsymbol{r}, \boldsymbol{r}') - g_0(\boldsymbol{r}, \boldsymbol{r}') \nabla^2 \psi_0(\boldsymbol{r})\right] dr \\
&= \iint_{\Omega_0} \left\{\psi_0(\boldsymbol{r}) \left[-\delta(\boldsymbol{r} - \boldsymbol{r}') - k_0^2 g_0(\boldsymbol{r}, \boldsymbol{r}')\right] + g_0(\boldsymbol{r}, \boldsymbol{r}') k_0^2 \psi_0(\boldsymbol{r})\right\} dr \\
&= -\iint_{\Omega_0} \psi_0(\boldsymbol{r}) \delta(\boldsymbol{r} - \boldsymbol{r}') dr \\
&= \begin{cases} -\psi_0(\boldsymbol{r}') \text{ if } \boldsymbol{r}' \in \Omega_0 \\ \quad 0 \text{ if } \boldsymbol{r}' \notin \Omega_0 \end{cases} \\
&= \int_{C_0} \left[\psi_0(\boldsymbol{r}) \boldsymbol{\nabla} g_0(\boldsymbol{r}, \boldsymbol{r}') - g_0(\boldsymbol{r}, \boldsymbol{r}') \boldsymbol{\nabla} \psi_0(\boldsymbol{r})\right] \cdot \boldsymbol{dS}, \qquad [1.16]
\end{aligned}$$

where the contour $C_0 = S \cup C_{0,\infty}$ bounding the medium Ω_0. In addition, the last line of equation [1.16] can be written as:

$$\begin{aligned}
&\int_{C_0} \left[\psi_0(\boldsymbol{r})\boldsymbol{\nabla} g_0(\boldsymbol{r},\boldsymbol{r}') - g_0(\boldsymbol{r},\boldsymbol{r}')\boldsymbol{\nabla}\psi_0(\boldsymbol{r})\right]\cdot \boldsymbol{dS} \\
&= -\int_S \left[\psi_0(\boldsymbol{r})\frac{\partial g_0(\boldsymbol{r},\boldsymbol{r}')}{\partial n} - g_0(\boldsymbol{r},\boldsymbol{r}')\frac{\partial \psi_0(\boldsymbol{r})}{\partial n}\right] dS \\
&\quad \underbrace{+\int_{C_{0,\infty}} \left[\psi_0(\boldsymbol{r})\frac{\partial g_0(\boldsymbol{r},\boldsymbol{r}')}{\partial n} - g_0(\boldsymbol{r},\boldsymbol{r}')\frac{\partial \psi_0(\boldsymbol{r})}{\partial n}\right] dS,}_{-\psi_{\text{inc}}(\boldsymbol{r}')\forall \boldsymbol{r}'\in\Omega_0}
\end{aligned} \quad [1.17]$$

where $\partial f/\partial n = \hat{\boldsymbol{n}} \cdot \boldsymbol{\nabla} f$. The minus sign before the integral over S is due to the sense of the normal $\hat{\boldsymbol{n}}$, which points inside Ω_0. The plus sign before the integral over $C_{0,\infty}$ is due to the sense of the normal $\hat{\boldsymbol{n}}_{0,\infty}$, which points outside Ω_0.

It can be shown that the integral over the contour $C_{0,\infty}$ equals minus the incident field $(-\psi_{\text{inc}}(\boldsymbol{r}'))$ [TSA 00]. For $\boldsymbol{r} \in S$, the substitution of equation [1.17] into equation [1.16] then leads to:

$$\begin{cases}
-\psi_{\text{inc}}(\boldsymbol{r}') = \int_S \left[\psi_0(\boldsymbol{r})\frac{\partial g_0(\boldsymbol{r},\boldsymbol{r}')}{\partial n} - g_0(\boldsymbol{r},\boldsymbol{r}')\frac{\partial \psi_0(\boldsymbol{r})}{\partial n}\right] dS & \text{if } \boldsymbol{r}' \notin \Omega_0 \\
\psi_0(\boldsymbol{r}') - \psi_{\text{inc}}(\boldsymbol{r}') = \int_S \left[\psi_0(\boldsymbol{r})\frac{\partial g_0(\boldsymbol{r},\boldsymbol{r}')}{\partial n} - g_0(\boldsymbol{r},\boldsymbol{r}')\frac{\partial \psi_0(\boldsymbol{r})}{\partial n}\right] dS & \text{if } \boldsymbol{r}' \in \Omega_0
\end{cases}. \quad [1.18]$$

Special attention must be paid when $\boldsymbol{r}'$ approaches the surface S from either below or above. For more details see [TSA 00 Chapter 4, section 2]. For $\boldsymbol{r}' \notin \Omega_0$, equation [1.18] gives the *extinction theorem*. For $\boldsymbol{r}' \in \Omega_0$ (S excluded), equation [1.18] gives the *Huygens' principle*:

$$\psi_{\text{sca},0}(\boldsymbol{r}') = \psi_0(\boldsymbol{r}') - \psi_{\text{inc}}(\boldsymbol{r}') = \int_S \left[\psi_0(\boldsymbol{r})\frac{\partial g_0(\boldsymbol{r},\boldsymbol{r}')}{\partial n} - g_0(\boldsymbol{r},\boldsymbol{r}')\frac{\partial \psi_0(\boldsymbol{r})}{\partial n}\right] dS. \quad [1.19]$$

This shows that the scattered field $\psi_{\text{sca},0}$ is expressed from the field ψ_0 and its normal derivative on the surface $\partial\psi_0/\partial n$. These two quantities are often called *surface currents*.

Applying the same principle for the field ψ_1 in the lower medium Ω_1, for $\boldsymbol{r} \in S$ we have [TSA 00]:

$$\begin{cases} 0 & = \int_S \left[\psi_1(\boldsymbol{r}) \dfrac{\partial g_1(\boldsymbol{r},\boldsymbol{r}')}{\partial n} - g_1(\boldsymbol{r},\boldsymbol{r}') \dfrac{\partial \psi_1(\boldsymbol{r})}{\partial n} \right] dS \quad \text{if } \boldsymbol{r}' \notin \Omega_1 \\ \psi_1(\boldsymbol{r}') & = - \int_S \left[\psi_1(\boldsymbol{r}) \dfrac{\partial g_1(\boldsymbol{r},\boldsymbol{r}')}{\partial n} - g_1(\boldsymbol{r},\boldsymbol{r}') \dfrac{\partial \psi_1(\boldsymbol{r})}{\partial n} \right] dS \quad \text{if } \boldsymbol{r}' \in \Omega_1 \end{cases}, \qquad [1.20]$$

and the scattered field $\psi_{\text{sca},1} = \psi_1$ since $\psi_{1,\text{inc}} = 0$ in Ω_1. In equations [1.18] and [1.20], the four surface unknowns to determine on the surface are $\psi_0(\boldsymbol{r})$, $\partial\psi_0(\boldsymbol{r})/\partial n$, $\psi_1(\boldsymbol{r})$ and $\partial\psi_1(\boldsymbol{r})/\partial n$, whereas the quantities $\psi_{\text{inc}}(\boldsymbol{r}')$, $g_0(\boldsymbol{r},\boldsymbol{r}')$ and $g_1(\boldsymbol{r},\boldsymbol{r}')$ are known. It is therefore necessary to have two additional equations, which are obtained from the boundary conditions [1.9] on the surface S, valid $\forall(\boldsymbol{r},\boldsymbol{r}') \in S$. Then, from equations [1.18] (the case for which is $\boldsymbol{r}' \in \Omega_0$), [1.20] (the case for which is $\boldsymbol{r}' \in \Omega_1$) and [1.9], $\forall(\boldsymbol{r},\boldsymbol{r}') \in S$, the IEs are [TSA 00]:

$$\begin{cases} \psi_{\text{inc}}(\boldsymbol{r}') = +\dfrac{1}{2}\psi_0(\boldsymbol{r}') - ⨍_S \psi_0(\boldsymbol{r}) \dfrac{\partial g_0(\boldsymbol{r},\boldsymbol{r}')}{\partial n} dS + \int_S g_0(\boldsymbol{r},\boldsymbol{r}') \dfrac{\partial \psi_0(\boldsymbol{r})}{\partial n} dS \\ 0 \quad = -\dfrac{1}{2}\psi_0(\boldsymbol{r}') - ⨍_S \psi_0(\boldsymbol{r}) \dfrac{\partial g_1(\boldsymbol{r},\boldsymbol{r}')}{\partial n} dS + \int_S \dfrac{1}{\rho_{01}} g_1(\boldsymbol{r},\boldsymbol{r}') \dfrac{\partial \psi_0(\boldsymbol{r})}{\partial n} dS \end{cases}. \qquad [1.21]$$

The symbol $⨍$ stands for the Cauchy principal value, which means that the case $\boldsymbol{r} = \boldsymbol{r}'$ is not accounted for in the calculation of the integral. In addition, letting $\int_S dS = ⨍ dS + \int_P dS$ (where P is a piece), it is important to note that for $\boldsymbol{r} = \boldsymbol{r}' \in S^+$ (at the upper limit), $\int_P \psi_0(\boldsymbol{r}) \frac{\partial g_0(\boldsymbol{r},\boldsymbol{r}')}{\partial n} dS = +\psi_0(\boldsymbol{r}')/2$, and for $\boldsymbol{r} = \boldsymbol{r}' \in S^-$ (at the lower limit), $\int_P \psi_0(\boldsymbol{r}) \frac{\partial g_0(\boldsymbol{r},\boldsymbol{r}')}{\partial n} dS = -\psi_0(\boldsymbol{r}')/2$.

To solve a scattering problem, the currents on the surfaces ψ_0 and $\partial\psi_0/\partial n$ must be calculated. For an arbitrary shape of a surface, a numerically rigorous method is needed to compute them because the IEs have no analytical solution. This is discussed in the following section. From the knowledge of these currents, the scattered field $\psi_{i,\text{sca}}$ ($i = \{1, 2\}$) is then computed in the domain $\Omega_i - S$ from equations [1.19] and [1.20] (with $\boldsymbol{r}' \in \Omega_i$), respectively.

COMMENT 1.1.– 2D SCALAR GREEN FUNCTION–. For a 2D problem, the Green function is expressed as:

$$g_i(\boldsymbol{r}, \boldsymbol{r}') = \frac{j}{4}\mathrm{H}_0^{(1)}(k_i\,\|\boldsymbol{r}-\boldsymbol{r}'\|) = \frac{j}{4}\mathrm{H}_0^{(1)}\left[k_i\sqrt{(x-x')^2+(z-z')^2}\right], \quad [1.22]$$

where $\mathrm{H}_0^{(1)}$ is the zeroth-order Hankel function of the first kind. Their derivatives with respect to x and z are then (with $D = \|\boldsymbol{r}-\boldsymbol{r}'\|$):

$$\left\{\begin{aligned} \frac{\partial g_i(\boldsymbol{r},\boldsymbol{r}')}{\partial x} &= \frac{\partial g_i(k_iD)}{\partial D}\frac{\partial D}{\partial x} = -\frac{jk_i}{4}\mathrm{H}_1^{(1)}(k_iD)\frac{x-x'}{D} \\ \frac{\partial g_i(\boldsymbol{r},\boldsymbol{r}')}{\partial z} &= \frac{\partial g_i(k_iD)}{\partial D}\frac{\partial D}{\partial z} = -\frac{jk_i}{4}\mathrm{H}_1^{(1)}(k_iD)\frac{z-z'}{D} \end{aligned}\right. . \quad [1.23]$$

The quantity $\partial g_i(\boldsymbol{r},\boldsymbol{r}')/\partial n = \hat{\boldsymbol{n}}\cdot\boldsymbol{\nabla} g_i(\boldsymbol{r},\boldsymbol{r}')$ is then:

$$\frac{\partial g_i(\boldsymbol{r},\boldsymbol{r}')}{\partial n} = -\frac{jk_i}{4}\frac{\mathrm{H}_1^{(1)}(k_i\,\|\boldsymbol{r}-\boldsymbol{r}'\|)}{\|\boldsymbol{r}-\boldsymbol{r}'\|}(\boldsymbol{r}-\boldsymbol{r}')\cdot\hat{\boldsymbol{n}}. \quad [1.24]$$

Figure 1.2 shows the real and the imaginary parts of $\mathrm{H}_0^{(1)}(x)$ and its envelope versus x ($x > 0$). For $x \gg 1$, it behaves as a periodic function because for $x \to +\infty$, we have [ABR 70]:

$$\mathrm{H}_0^{(1)}(x) \approx \sqrt{\frac{2}{x\pi}}\exp\left[j\left(x-\frac{\pi}{4}\right)\right] \quad \text{as } x \to +\infty. \quad [1.25]$$

Moreover, the equation of its envelope is $\sqrt{2/(x\pi)} \propto \sqrt{1/x}$, which corresponds to a *cylindrical* wave (2D problem).

1.2.4. *Radar cross-section (RCS)*

For a 2D problem, in medium Ω_0, the RCS is defined as:

$$\mathrm{RCS} = \lim_{r'\to\infty} 2\pi r'\left|\frac{\psi_{\mathrm{sca},0}}{\psi_{\mathrm{inc},0}}\right|^2, \quad [1.26]$$

where r' is the distance between the object (phase center) and the receiver.

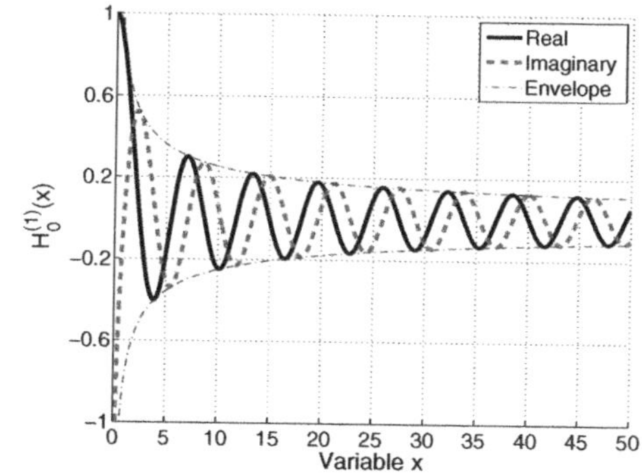

Figure 1.2. *Real and the imaginary parts of $H_0^{(1)}(x)$ and its envelope versus x ($x > 0$)*

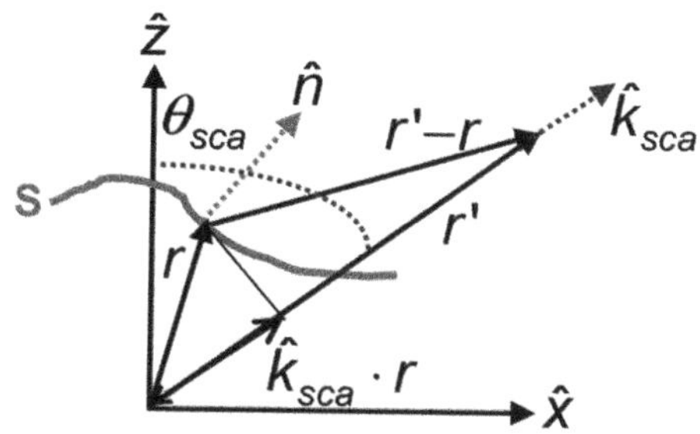

Figure 1.3. *Huygens' principle in the far field for a 2D problem*

As shown in Figure 1.3, in the far field ($r' \to \infty$, $r' \gg$ and $r' \gg \lambda_0 = 2\pi/k_0$), $\|\boldsymbol{r} - \boldsymbol{r}'\| = \|\boldsymbol{r}' - \boldsymbol{r}\| \approx r' - \hat{\boldsymbol{k}}_{\text{sca}} \cdot \boldsymbol{r}$ for the phase, whereas for the amplitude term $\|\boldsymbol{r} - \boldsymbol{r}'\| \approx r'$. From equation [1.25], in the observation direction defined by $\hat{\boldsymbol{k}}_{\text{sca}}$, the Green function can be simplified in the far field as:

$$g_0(\boldsymbol{r}, \boldsymbol{r}') = \frac{j}{4} \mathrm{H}_0^{(1)}(k_0 \|\boldsymbol{r} - \boldsymbol{r}'\|) \approx \frac{j}{4} \sqrt{\frac{2}{\pi k_0 r'}} e^{j(k_0 r' - \boldsymbol{k}_{\text{sca}} \cdot \boldsymbol{r} - \pi/4)}. \quad [1.27]$$

Moreover,

$$\frac{\partial g_0(\boldsymbol{r}, \boldsymbol{r}')}{\partial n} = \hat{\boldsymbol{n}} \cdot \nabla_{\boldsymbol{r}'} g_0(\boldsymbol{r}, \boldsymbol{r}') \approx -j g_0(\boldsymbol{r}, \boldsymbol{r}') \boldsymbol{k}_{\text{sca}} \cdot \hat{\boldsymbol{n}}. \quad [1.28]$$

Substituting these two equations into the Huygens' principle [1.19], the scattered field in medium Ω_0 in the far-field zone (super script ∞) is:

$$\psi_{\text{sca},0}^{\infty} = \frac{j}{4}\sqrt{\frac{2}{\pi k_0 r'}} e^{-j\pi/4 + k_0 r'} \psi_{\text{sca},0}^{\infty,0} \psi_{\text{inc},0}, \qquad [1.29]$$

where the variable $\psi_{\text{sca},0}^{\infty,0}$ is:

$$\psi_{\text{sca},0}^{\infty,0} = -\frac{1}{\psi_{\text{inc},0}} \int_S \underbrace{\left[j\boldsymbol{k}_{\text{sca}} \cdot \hat{\boldsymbol{n}} \psi_0(\boldsymbol{r}) + \frac{\partial \psi_0(\boldsymbol{r})}{\partial n} \right]}_{f(\boldsymbol{r})} e^{-j\boldsymbol{k}_{\text{sca}} \cdot \boldsymbol{r}} dS, \qquad [1.30]$$

and $\psi_{\text{inc},0}$ is the modulus of the incident field ψ_{inc} in medium Ω_0. Then, the scattered far field is then obtained from the Fourier transform of $f(\boldsymbol{r})$.

The substitution of equation [1.29] into equation [1.26] then leads to:

$$\text{RCS} = \frac{\left|\psi_{\text{sca},0}^{\infty,0}\right|^2}{4\,|k_0|}. \qquad [1.31]$$

Since $\psi_{\text{sca},0}^{\infty,0}$ is dimensionless, the RCS has the dimension of a distance (meter), since we consider it to be a 2D problem.

1.2.5. *Normalized radar cross-section (NRCS)*

If the surface has a finite extent, edge diffraction occurs, because the incident field does not vanish on the edges of the surface. To reduce this phenomenon, for the scattering from a rough surface, a tapered incident wave is used instead of a plane incident wave. A possible option is the Thorsos wave defined as [THO 88]:

$$\psi_{\text{inc}}(\boldsymbol{r}) = \psi_{\text{inc},0} \underbrace{\exp\left(j\boldsymbol{k}_{\text{inc}} \cdot \boldsymbol{r}\right)}_{\text{Plane wave}} \underbrace{\exp\left(-\frac{(x + z\tan\theta_{\text{inc}})^2}{g^2}\right)}_{\text{Damping factor}} \underbrace{\exp\left[jw(\boldsymbol{r})\boldsymbol{k}_{\text{inc}} \cdot \boldsymbol{r}\right]}_{\text{Corrective term}},$$

[1.32]

where

$$w(\boldsymbol{r}) = \left[\frac{2(x + z\tan\theta_{\text{inc}})^2}{g^2} - 1\right]\frac{1}{(k_0 g\cos\theta_{\text{inc}})^2}, \qquad [1.33]$$

and $\boldsymbol{k}_{\text{inc}} = k_0(\sin\theta_i\hat{\boldsymbol{x}} - \cos\theta_i\hat{\boldsymbol{z}})$. The damping is orthogonal to the incident vector $\boldsymbol{k}_{\text{inc}}$. The additional corrective term allows us to better satisfy the Helmholtz equation. Nevertheless, the Thorsos wave verifies the Helmholtz equation at the order $\mathcal{O}\left(\frac{1}{(gk_0\cos\theta_{\text{inc}})^3}\right)$, implying that $1/C = gk_0\cos\theta_{\text{inc}} >> 1$. This condition is not satisfied for:

– grazing incidence angles: Indeed, for given k_0 and g, if $\theta_{\text{inc}} \to \pi/2$, then $gk_0\cos\theta_{\text{inc}} \to 0$.

– g small in comparison to the wavelength λ_0: in other words, if the width of the incident beam is small in comparison to the wavelength.

Typically, $1/(gk_0\cos\theta_{\text{inc}}) \leq C = 0.037$ and in the following, we take $g = L/6$, where L is the surface length.

From the Thorsos wave, the normalized incident power on the rough surface mean plane $z = 0$ is then:

$$\begin{aligned}
p_{\text{inc}} &= -\frac{1}{|\psi_{\text{inc},0}|^2}\int_{-\infty}^{+\infty} \boldsymbol{S}_{\text{inc}}\cdot\hat{\boldsymbol{z}}\big|_{z=0}\,dx \\
&= -\frac{1}{2|\psi_{\text{inc},0}|^2\eta_0}\int_{-\infty}^{+\infty} |\psi_{\text{inc}}|^2\hat{\boldsymbol{k}}_{\text{inc}}\cdot\hat{\boldsymbol{z}}\Big|_{z=0}\,dx \\
&= \frac{\cos\theta_{\text{inc}}}{2|\psi_{\text{inc},0}|^2\eta_0}\int_{-\infty}^{+\infty} |\psi_{\text{inc}}|^2\big|_{z=0}\,dx \\
&= \frac{g\cos\theta_{\text{inc}}}{2\eta_0}\sqrt{\frac{\pi}{2}}\left[1 - \frac{1+2\tan^2\theta_{\text{inc}}}{2k_0^2g^2\cos^2\theta_{\text{inc}}}\right],
\end{aligned} \qquad [1.34]$$

where η_0 is the wave impedance in medium Ω_0. In addition, $\boldsymbol{S}_{\text{inc}}$ is the Poynting vector that gives the power density carried by the incident wave. The NRCS is defined as [TSA 00]:

$$\text{NRCS} = \lim_{r'\to\infty}\frac{r'}{2\eta_0}\frac{\left|\psi_{\text{sca},0}^{\infty}\right|^2}{p_{\text{inc}}} = \frac{1}{16\pi\eta_0 k_0}\frac{\left|\psi_{\text{sca},0}^{\infty,0}\right|^2}{p_{\text{inc}}}, \qquad [1.35]$$

where the field $\psi_{\text{sca},0}^{\infty,0}$ is computed from equation [1.30]. Note that unlike the RCS, the NRCS is dimensionless.

1.3. Method of moments with point-matching method

The MoM is a numerical method that has been used extensively for the solution of scattering electromagnetic problems. Many excellent textbooks like [HAR 68, TSA 00] have been written on this subject. A characteristic of this technique is that it leads to a full matrix equation that can be solved from a matrix inversion. In this book, the MoM with point-matching method and pulse basis function is applied. Their main advantages are that they are simple to program and are efficient for scattering from rough surfaces.

Consider a 1D IE of the form:

$$\mathcal{L}(f) = g, \tag{1.36}$$

where $\mathcal{L}$ is an integral operator or integral–differential operator, f is the unknown function and g is a given function.

– *Step 1: Basis functions*

A set of N basis functions in the domain $\mathcal{D}$ is chosen. Let the basis functions be $f_1, f_2, \ldots, f_N$. The unknown function is expanded in terms of a linear combination of these functions:

$$f \simeq \tilde{f} = \sum_{n=1}^{N} a_n f_n, \tag{1.37}$$

and $\tilde{f}$ verifies $\lim_{N\to+\infty} \left|f - \tilde{f}\right| = 0$. The substitution of equation [1.37] into equation [1.36] leads to:

$$\mathcal{L}f = \mathcal{L}\left(\sum_{n=1}^{N} a_n f_n\right) + \varepsilon = g, \tag{1.38}$$

where ε is the residue due to the truncation of the sum at the order N. Since $\mathcal{L}f$ is a linear operator:

$$\mathcal{L}f = \sum_{n=1}^{N} a_n(\mathcal{L}f_n) + \varepsilon = g. \tag{1.39}$$

– *Step 2: Testing functions*

Next, a set of M weighting functions $w_1, w_2, \ldots, w_M$ is chosen. Multiplying equation [1.39] by $w_m(x)$ (with $m = 1 \ldots M$), assuming that $\langle w_m, \varepsilon \rangle \approx 0$ and integrating over the domain $\mathcal{D}$, we obtain:

$$\left\langle w_m, \sum_{n=1}^{N} a_n(\mathcal{L}f_n) \right\rangle = \sum_{n=1}^{N} a_n \langle w_m, \mathcal{L}f_n \rangle = \langle w_m, g \rangle\,, \qquad [1.40]$$

where the inner product $\langle \ldots \rangle$ is defined for a single variable x as:

$$\langle f, g \rangle = \int_{\mathcal{D}} f(x)g(x)dx. \qquad [1.41]$$

– *Step 3: Linear system*

From equation [1.40], the linear system to be solved is:

$$\bar{\boldsymbol{Z}}\boldsymbol{X} = \boldsymbol{b}, \qquad [1.42]$$

in which the elements of the matrix $\bar{\boldsymbol{Z}}$ and the vector $\boldsymbol{b}$ are defined as:

$$\begin{cases} Z_{mn} = \langle w_m, \mathcal{L}f_n \rangle \\ b_m = \langle w_m, g \rangle \end{cases}, \qquad [1.43]$$

and the elements of the vector $\boldsymbol{X}$, which equals a_n, must be determined. The matrix $\bar{\boldsymbol{Z}}$ is the impedance matrix of the scattering problem and depends on the shape and the electric properties of the surface.

– *Point-matching method*

Basis functions can use full domain functions such as sines, cosines, special functions, polynomials and modal solutions. A set that is useful for a practical problem is the subsectional basis function. This means that each f_n is only non-zero over a subsection of the domain of f.

A common choice is the pulse function:

$$f_n(x) = \begin{cases} 1 \text{ if } \alpha_n \leq x \leq \beta_n \\ 0 \text{ otherwise} \end{cases}, \qquad [1.44]$$

where the interval $\mathcal{D}$ is divided into N subintervals $\beta_n - \alpha_n$ with end points α_1 and β_N with $n = \{1, 2, \ldots, N\}$.

For the weighting functions w_m, the following are two common choices:

1) Galerkin's method. In this case, the weighting functions are the same as the basis function, that is $w_n(x) = f_n(x)$ with $n = \{1, 2, \ldots, N\}$.

2) Point-matching method. In this case, the weighting functions are the Dirac delta functions $w_m = \delta(x - x_m)$ with $m = \{1, 2, \ldots, M\}$, and we choose $M = N$.

From the point-matching method and equation [1.40], we obtain:

$$\begin{aligned} Z_{mn} &= \langle w_m(x), \mathcal{L}[f_n(x)] \rangle = \langle \delta(x - x_m), \mathcal{L}[f_n(x)] \rangle \\ &= \int_{\mathcal{D}} \delta(x - x_m) \mathcal{L}[f_n(x)] dx = \mathcal{L}[f_n(x_m)], \end{aligned} \quad [1.45]$$

and

$$\begin{aligned} b_m &= \langle w_m, g(x) \rangle = \langle \delta(x - x_m), g(x) \rangle \\ &= \int_{\mathcal{D}} \delta(x - x_m) g(x) dx = g(x_m). \end{aligned} \quad [1.46]$$

1.4. Application to a surface

In this section, the MoM is applied along with the point-matching method to convert the IEs [1.21] into a linear system.

1.4.1. *The Dirichlet boundary conditions*

For a PC surface and for the TE polarization (the Dirichlet boundary conditions), the field vanishes on the surface, $\psi_0 = 0$. Thus, equation [1.21] can be simplified as:

$$\psi_{\text{inc}}(\boldsymbol{r}') = \int_S g_0(\boldsymbol{r}, \boldsymbol{r}') \frac{\partial \psi_0(\boldsymbol{r})}{\partial n} dS, \quad [1.47]$$

where the points $\boldsymbol{r} = (x, y)$ and $\boldsymbol{r}' = (x', y')$ are on the surface S.

For a surface of length L and centered on $x = 0$, equation [1.47] can be written as:

$$\mathcal{L}f = g, \tag{1.48}$$

where

$$\begin{cases} \mathcal{L}\bullet = \displaystyle\int_{-L/2}^{L/2} \sqrt{1 + \left(\frac{dz}{dx}\right)^2} g_0(\boldsymbol{r}, \boldsymbol{r}')dx\bullet \\ f = \frac{\partial\psi(\boldsymbol{r})}{\partial n}, \quad g = \psi_{\text{inc}}(\boldsymbol{r}') \end{cases}. \tag{1.49}$$

The unknown is f. From the MoM, equation [1.47] is converted into a linear system $\bar{\boldsymbol{Z}}\boldsymbol{X} = \boldsymbol{b}$. The elements of the impedance matrix $\bar{\boldsymbol{Z}}$, Z_{mn} (with $(n, m) \in [1; N]$), and the components of the vector $\boldsymbol{b}$, b_m, are given from equations [1.45] and [1.46] by:

$$\begin{aligned} Z_{mn} &= \int_{\alpha_n}^{\beta_n} \sqrt{1 + \left(\frac{dz}{dx}\right)^2} g_0(\boldsymbol{r}, \boldsymbol{r}_m)dx \\ &\approx \sqrt{1 + \left(\frac{dz_n}{dx_n}\right)^2} g_0(\boldsymbol{r}_n, \boldsymbol{r}_m)(\beta_n - \alpha_n), \end{aligned} \tag{1.50}$$

and

$$b_m = \psi_{\text{inc}}(\boldsymbol{r}_m). \tag{1.51}$$

For the calculation of the integration over x of Z_{mn}, we assumed that the integrand does not vary significantly on the range $x \in [\alpha_n; \beta_n]$. Physically, this condition is fulfilled if $\beta_n - \alpha_n \ll \lambda_0$ (corresponding to slow variations of the Green function g_0 over the distance $\beta_n - \alpha_n$), where λ_0 is the wave length in medium Ω_0. Typically, for the simulations $\beta_n - \alpha_n = \lambda_0/10$, corresponding to a distance for which the Green function slowly varies. For $m = n$, the Green function has a singularity. Then, the evaluation of integral [1.50] requires us to make Taylor series expansions of the integrand around $\boldsymbol{r} = \boldsymbol{r}_m$. For more details, see [TSA 00, Chapter 4]. In conclusion, the elements of the impedance matrix are:

$$Z_{mn} = \frac{j\Delta_n\sqrt{1+\gamma_n^2}}{4}\begin{cases}\left[1+\frac{2j}{\pi}\ln\left(0.164k_0\sqrt{1+\gamma_n^2}\Delta_n\right)\right] & \text{for } m=n \\ \mathrm{H}_0^{(1)}(k_0\left\|\boldsymbol{r}_n-\boldsymbol{r}_m\right\|) & \text{for } m\neq n\end{cases}, \quad [1.52]$$

where $\|\boldsymbol{r}_n - \boldsymbol{r}_m\| = \sqrt{(x_n - x_m)^2 + (z_n - z_m)^2}$

and
$$\begin{cases} X_n = \dfrac{\partial\psi(\boldsymbol{r}_n)}{\partial n} \\ b_m = \psi_{\text{inc}}(\boldsymbol{r}_m)\end{cases}, \quad [1.53]$$

with $\gamma_n = dz_n/dx_n$ and $\Delta_n = \beta_n - \alpha_n$.

1.4.2. *The Neumann boundary conditions*

For a PC surface and for the TM polarization (the Neumann boundary conditions), the normal derivative of the field vanishes on the surface, $\partial\psi_0/\partial n = 0$. Thus, equation [1.21] can be simplified as:

$$\psi_{\text{inc}}(\boldsymbol{r}') = \frac{1}{2}\psi_0(\boldsymbol{r}') - \int_S \psi_0(\boldsymbol{r})\frac{\partial g_0(\boldsymbol{r},\boldsymbol{r}')}{\partial n}dS, \quad [1.54]$$

where the points $\boldsymbol{r} = (x, y)$ and $\boldsymbol{r}' = (x', y')$ are on the surface.

Using the same method as in the previous section and from equation [1.24], we have:

$$Z_{mn} = \begin{cases} -\dfrac{jk_0\Delta_n}{4}\dfrac{\mathrm{H}_1^{(1)}(k_0\|\boldsymbol{r}_n-\boldsymbol{r}_m\|)}{\|\boldsymbol{r}_n-\boldsymbol{r}_m\|} \\ \qquad\qquad \times\left[\gamma_n(x_n-x_m)-(z_n-z_m)\right] & \text{for } m\neq n, \\ +\dfrac{1}{2}-\dfrac{\Delta_n}{4\pi}\dfrac{\gamma_n'}{1+\gamma_n^2} & \text{for } m=n\end{cases} \quad [1.55]$$

where $\gamma'_n = \frac{d\gamma_n}{dx_n}$ and

$$\left\{\begin{array}{l} X_n = \psi(\boldsymbol{r}_n) \\ b_m = \psi_{\text{inc}}(\boldsymbol{r}_m). \end{array}\right. \qquad [1.56]$$

1.4.3. *General case*

For a dielectric medium Ω_1, equation [1.21] leads to:

$$\left\{\begin{array}{ll} \psi_{\text{inc}}(\boldsymbol{r}') & = \underbrace{+\frac{1}{2}\psi_0(\boldsymbol{r}') - \fint_S \psi_0(\boldsymbol{r})\frac{\partial g_0(\boldsymbol{r},\boldsymbol{r}')}{\partial n}dS}_{\text{Neumann}} + \underbrace{\int_S g_0(\boldsymbol{r},\boldsymbol{r}')\frac{\partial \psi_0(\boldsymbol{r})}{\partial n}dS}_{\text{Dirichlet}} \\ 0 & = \underbrace{-\frac{1}{2}\psi_0(\boldsymbol{r}') - \fint_S \psi_0(\boldsymbol{r})\frac{\partial g_1(\boldsymbol{r},\boldsymbol{r}')}{\partial n}dS}_{\text{Neumann with } k_0 \to k_1} + \underbrace{\int_S \frac{1}{\rho_{01}} g_1(\boldsymbol{r},\boldsymbol{r}')\frac{\partial \psi_0(\boldsymbol{r})}{\partial n}dS}_{\text{Dirichlet with } k_0 \to k_1} \end{array}\right. . \qquad [1.57]$$

The general case is then obtained from "linear combinations" of the Neumann and Dirichlet boundary conditions. The discretization of the above equations from the MoM then leads to the following impedance matrix:

$$\bar{\boldsymbol{Z}} = \begin{bmatrix} \bar{\boldsymbol{Z}}_{\text{Neu}} & \bar{\boldsymbol{Z}}_{\text{Dir}} \\ \bar{\boldsymbol{Z}}_{\text{Neu},k_0\to k_1} & \frac{1}{\rho_{01}}\bar{\boldsymbol{Z}}_{\text{Dir},k_0\to k_1} \end{bmatrix}, \qquad [1.58]$$

where the subscripts "Neu" and "Dir" mean Neumann and Dirichlet, respectively. It is important to note that for $\bar{\boldsymbol{Z}}_{\text{Neu},k_0\to k_1}$, the Cauchy principal value is $-1/2$, instead of $+1/2$ for $\bar{\boldsymbol{Z}}_{\text{Neu}}$. In addition, the vectors $\boldsymbol{b}$ and $\boldsymbol{X}$ are:

$$\boldsymbol{b} = \begin{bmatrix} \psi_{\text{inc}}(\boldsymbol{r}_1) \\ \psi_{\text{inc}}(\boldsymbol{r}_2) \\ \vdots \\ \psi_{\text{inc}}(\boldsymbol{r}_N) \\ \left.\begin{matrix} 0 \\ 0 \\ \vdots \\ 0 \end{matrix}\right\} N \text{ times} \end{bmatrix} \quad \text{and} \quad \mathbf{X} = \begin{bmatrix} \psi_0(\boldsymbol{r}_1) \\ \psi_0(\boldsymbol{r}_2) \\ \vdots \\ \psi_0(\boldsymbol{r}_N) \\ \frac{\partial\psi_0(\boldsymbol{r}_1)}{\partial n} \\ \frac{\partial\psi_0(\boldsymbol{r}_2)}{\partial n} \\ \vdots \\ \frac{\partial\psi_0(\boldsymbol{r}_N)}{\partial n} \end{bmatrix}. \qquad [1.59]$$

The square matrix $\bar{Z}$ is of size $2N \times 2N$. The square matrices $\bar{Z}_{\text{Dir}}$ and $\bar{Z}_{\text{Neu}}$ of size $N \times N$ are expressed from equations [1.52] and [1.55], respectively.

For a dielectric media Ω_1, the sampling step $\beta_n - \alpha_n$ is $(\lambda_0/10)/|\sqrt{\epsilon_{r1}}|$ instead of $\lambda_0/10$ for a PC surface because the wave number in medium Ω_1 is $k_1 = k_0\sqrt{\epsilon_{r1}}$, where k_0 is the wave number in a vacuum. Thus, for a highly conducting surface checking $\text{Im}(\epsilon_{r1}) \gg 1$, the number of samples on the surface increases significantly, which makes the inversion of the impedance matrix very difficult. To overcome this issue, the impedance boundary condition (IBC) is applied.

1.4.4. *Impedance boundary condition*

For $\text{Im}(\epsilon_{r1}) \gg 1$, system [1.57] can be converted into only one IES. Indeed, from the Leontovitch boundary condition which is also called IBC, we have on the surface $\boldsymbol{r} \in S$:

$$\begin{cases} \text{TE} : \psi_0(\boldsymbol{r}) = \dfrac{j}{k_0}\sqrt{\dfrac{\epsilon_{r0}}{\epsilon_{r1}}}\dfrac{\partial\psi_0(\boldsymbol{r})}{\partial n} \\ \text{TM} : \dfrac{\partial\psi_0(\boldsymbol{r})}{\partial n} = \dfrac{k_0}{j}\sqrt{\dfrac{\epsilon_{r0}}{\epsilon_{r1}}}\psi_0(\boldsymbol{r}) \end{cases}, \qquad [1.60]$$

where $\text{Im}(\epsilon_{r1}) \gg 1$. System [1.57] is then simplified as:

$$\begin{cases} \text{TE} : \psi_{\text{inc}}(\boldsymbol{r}') = \displaystyle\int_S \frac{\partial\psi_0(\boldsymbol{r})}{\partial n}\left[g_0(\boldsymbol{r},\boldsymbol{r}') - \frac{j}{k_0}\sqrt{\frac{\epsilon_{r0}}{\epsilon_{r1}}}\frac{\partial g_0(\boldsymbol{r},\boldsymbol{r}')}{\partial n}\right] dS \\ \text{TM} : \psi_{\text{inc}}(\boldsymbol{r}') = \dfrac{1}{2}\psi_0(\boldsymbol{r}') + \displaystyle\fint_S \psi_0(\boldsymbol{r})\left[g_0(\boldsymbol{r},\boldsymbol{r}')\frac{k_0}{j}\sqrt{\frac{\epsilon_{r0}}{\epsilon_{r1}}} - \frac{\partial g_0(\boldsymbol{r},\boldsymbol{r}')}{\partial n}\right] dS \end{cases}. \qquad [1.61]$$

For the TE polarization, it is important to note that the unknown is $\partial\psi_0(\boldsymbol{r})/\partial n$ since $|\partial\psi_0(\boldsymbol{r})/\partial n| \gg |\psi_0(\boldsymbol{r})|$, which is similar to considering the Dirichlet boundary condition. On the other hand, for the TM polarization, the unknown is $\psi_0(\boldsymbol{r})$ since $|\psi_0(\boldsymbol{r})| \gg |\partial\psi_0(\boldsymbol{r})/\partial n|$, which is similar to considering the Neumann boundary condition.

From the MoM, the matrix impedance is then:

$$\bar{\boldsymbol{Z}} = \beta \bar{\boldsymbol{Z}}_{\text{Dir}} + \alpha \bar{\boldsymbol{Z}}_{\text{Neu}}, \qquad [1.62]$$

with

$$\begin{cases} \text{TE} : \beta = 1, & \alpha = \dfrac{j}{k_0}\sqrt{\dfrac{\epsilon_{r0}}{\epsilon_{r1}}},\ \boldsymbol{X} = \dfrac{\partial \psi_0(\boldsymbol{r})}{\partial n},\ \psi_0(\boldsymbol{r}) = \alpha \dfrac{\partial \psi_0(\boldsymbol{r})}{\partial n} \\ \text{TM} : \beta = \dfrac{k_0}{j}\sqrt{\dfrac{\epsilon_{r0}}{\epsilon_{r1}}},\ \alpha = 1, & \boldsymbol{X} = \psi_0(\boldsymbol{r}),\ \dfrac{\partial \psi_0(\boldsymbol{r})}{\partial n} = \beta \psi_0(\boldsymbol{r}) \end{cases}. \qquad [1.63]$$

In conclusion, the use of the IBC allows us to discretize the surface along the wave number k_0 and then it becomes independent of the permittivity of medium Ω_1. In terms of number of unknowns, it is equivalent to solving the scattering by a PC surface.

1.5. Forward–Backward (FB) method

For a problem with many unknowns, it is interesting to investigate rigorous fast numerical methods to treat the scattering from a large electrically rough surface. For instance, for a single rough surface, we can quote the banded-matrix-iterative-approach/canonical grid (BMIA-CAG) of Tsang *et al.* [TSA 93a, TSA 93b, TSA 95] of complexity $\mathcal{O}(N \log N)$, the FB method of Holliday *et al.* [KAP 96, ADA 96, HOL 98, IOD 02] of complexity $\mathcal{O}(N^2)$ and the accelerated version FB method with spectral acceleration (FB-SA) of Chou *et al.* [CHO 02, CHO 00, TOR 00, TOR 02] of complexity $\mathcal{O}(N)$, in which N is the number of unknowns on the surface.

In this section, the FB method is applied to a dielectric surface to speed up the calculation of $\bar{\boldsymbol{Z}}^{-1}\boldsymbol{b}$, in order to reduce the complexity to $\mathcal{O}(N^2)$ instead of $\mathcal{O}(N^3)$ from a direct lower–upper (LU) inversion.

We want to solve $\bar{\boldsymbol{Z}}\boldsymbol{X} = \boldsymbol{b} \Leftrightarrow \boldsymbol{X} = \bar{\boldsymbol{Z}}^{-1}\boldsymbol{b}$. From equation [1.58], the matrix $\bar{\boldsymbol{Z}}$ of size $2N \times 2N$ can be expressed from four square submatrices of size $N \times N$ as:

$$\bar{\boldsymbol{Z}} = \begin{bmatrix} \bar{\boldsymbol{A}} & \bar{\boldsymbol{B}} \\ \bar{\boldsymbol{C}} & \bar{\boldsymbol{D}} \end{bmatrix},\ \boldsymbol{X} = \begin{bmatrix} \boldsymbol{X}_1 \\ \boldsymbol{X}_2 \end{bmatrix},\ \boldsymbol{b} = \begin{bmatrix} \boldsymbol{b}_1 \\ \boldsymbol{b}_2 \end{bmatrix}. \qquad [1.64]$$

The FB algorithm decomposes $\bar{\boldsymbol{Z}}\boldsymbol{X} = \boldsymbol{b}$ into:

$$\begin{cases} \bar{\boldsymbol{A}}_{\text{Diag}}\boldsymbol{X}_{1,\text{Forw}} + \bar{\boldsymbol{B}}_{\text{Diag}}\boldsymbol{X}_{2,\text{Forw}} = \boldsymbol{b}_1 - \bar{\boldsymbol{A}}_{\text{Forw}}\boldsymbol{X}_1 - \bar{\boldsymbol{B}}_{\text{Forw}}\boldsymbol{X}_2 \\ \bar{\boldsymbol{C}}_{\text{Diag}}\boldsymbol{X}_{1,\text{Forw}} + \bar{\boldsymbol{D}}_{\text{Diag}}\boldsymbol{X}_{2,\text{Forw}} = \boldsymbol{b}_2 - \bar{\boldsymbol{C}}_{\text{Forw}}\boldsymbol{X}_1 - \bar{\boldsymbol{D}}_{\text{Forw}}\boldsymbol{X}_2 \end{cases}, \quad [1.65]$$

and

$$\begin{cases} \bar{\boldsymbol{A}}_{\text{Diag}}\boldsymbol{X}_{1,\text{Back}} + \bar{\boldsymbol{B}}_{\text{Diag}}\boldsymbol{X}_{2,\text{Back}} = -\bar{\boldsymbol{A}}_{\text{Back}}\boldsymbol{X}_1 - \bar{\boldsymbol{B}}_{\text{Back}}\boldsymbol{X}_2 \\ \bar{\boldsymbol{C}}_{\text{Diag}}\boldsymbol{X}_{1,\text{Back}} + \bar{\boldsymbol{D}}_{\text{Diag}}\boldsymbol{X}_{2,\text{Back}} = -\bar{\boldsymbol{C}}_{\text{Back}}\boldsymbol{X}_1 - \bar{\boldsymbol{D}}_{\text{Back}}\boldsymbol{X}_2 \end{cases}. \quad [1.66]$$

For instance, $\bar{\boldsymbol{A}}_{\text{Diag}}$ is a *diagonal* matrix, $\bar{\boldsymbol{A}}_{\text{Forw}}$ a *lower* triangular matrix and $\bar{\boldsymbol{A}}_{\text{Back}}$ an *upper* triangular matrix, all built from $\bar{\boldsymbol{A}}$ ($\bar{\boldsymbol{A}} = \bar{\boldsymbol{A}}_{\text{Forw}} + \bar{\boldsymbol{A}}_{\text{Diag}} + \bar{\boldsymbol{A}}_{\text{Back}}$). The subscripts $\{\text{Diag}, \text{Forw}, \text{Back}\}$ stand for diagonal, forward and backward matrices and are referred to as diagonal, lower and upper triangular matrices, respectively. Moreover, $\{\boldsymbol{b}_1, \boldsymbol{b}_2, \boldsymbol{X}_1, \boldsymbol{X}_2\}$ are column vectors of length N. Finally, the unknown vectors are decomposed into $\boldsymbol{X}_i = \boldsymbol{X}_{i,\text{Forw}} + \boldsymbol{X}_{i,\text{Back}}$ ($i = \{1, 2\}$), in which $\boldsymbol{X}_{i,\text{Forw}}$ gives the *forward* contribution (from the points on the *left* of the current point) and $\boldsymbol{X}_{i,\text{Back}}$ gives the *backward* contribution (from the *right*). The surface is oriented by assuming that the incident beam propagates from the left to the right.

To compute $\boldsymbol{X}_i = \boldsymbol{X}_{i,\text{Forw}} + \boldsymbol{X}_{i,\text{Back}}$, an iterative procedure is applied. Assuming first that $\boldsymbol{X}_{i,\text{Back}} = \boldsymbol{0} \Rightarrow \boldsymbol{X}_i = \boldsymbol{X}_{i,\text{Forw}} + \boldsymbol{X}_{i,\text{Back}} = \boldsymbol{X}_{i,\text{Forw}}$, equation [1.65] is solved for $\boldsymbol{X}_{i,\text{Forw}}$. Then, by introducing $\boldsymbol{X}_{i,\text{Forw}}$ in equation [1.66], we obtain $\boldsymbol{X}_{i,\text{Back}}$. The first iteration $\boldsymbol{X}_i^{(0)}$ is then equal to $\boldsymbol{X}_{i,\text{Forw}} + \boldsymbol{X}_{i,\text{Back}}$. The scheme is repeated to calculate the next iterations $\boldsymbol{X}_i^{(p)}$ up to the order $p = P_{\text{FB}}$.

The use of equations [1.65] and [1.66] is very convenient to solve by substitution for $\boldsymbol{X}_{i,\text{Forw}}$ and $\boldsymbol{X}_{i,\text{Back}}$. For instance, from [1.65], since $\{\bar{\boldsymbol{A}}_{\text{Forw}}, \bar{\boldsymbol{B}}_{\text{Forw}}, \bar{\boldsymbol{C}}_{\text{Forw}}, \bar{\boldsymbol{D}}_{\text{Forw}}\}$ are lower triangular matrices with null diagonal coefficients, we get with $m \in [2; N]$:

$$\begin{cases} A_{\text{Diag}}^{m,m} X_{1,\text{Forw}}^m + B_{\text{Diag}}^{m,m} X_{2,\text{Forw}}^m = b_1^m - \sum\limits_{n=1}^{n=m-1} \left(A_{\text{Forw}}^{m,n} X_1^n + B_{\text{Forw}}^{m,n} X_2^n\right) \\ C_{\text{Diag}}^{m,m} X_{1,\text{Forw}}^m + D_{\text{Diag}}^{m,m} X_{2,\text{Forw}}^m = b_2^m - \sum\limits_{n=1}^{n=m-1} \left(C_{\text{Forw}}^{m,n} X_1^n + D_{\text{Forw}}^{m,n} X_2^n\right) \end{cases}. \quad [1.67]$$

For instance, $A^{m,n}$ is the element of the matrix $\bar{\boldsymbol{A}}$ of column m and row n. X_i^n is the nth component of the vector $\boldsymbol{X}_i$. Thus, assuming first $\boldsymbol{X}_{\text{Back}} = \boldsymbol{0} \Rightarrow \boldsymbol{X}_i = \boldsymbol{X}_{i,\text{Forw}} + \boldsymbol{X}_{i,\text{Back}} = \boldsymbol{X}_{i,\text{Forw}}$ and by solving equation [1.67], the unknowns $\{X_{1,\text{Forw}}^m, X_{2,\text{Forw}}^m\}$ with $m \in [2; N]$ are calculated from $4N^2/2$ multiplications. From equation [1.66], we obtain a similar equation system to [1.67] but the sum over n is $n \in [m+1; N]$, and the unknowns $\{X_{1,\text{Back}}^m, X_{1,\text{Back}}^m\}$ with $m \in [1; N-1]$ are also calculated from $4N^2/2$ multiplications. In conclusion, the complexity of the FB method is $\mathcal{O}(N^2)$.

Déchamps *et al.* [DÉC 07b] mathematically showed that the FB method converges if the spectral radius (i.e. the modulus of the eigenvalue, which has the highest modulus) of the characteristic matrix $\bar{\boldsymbol{M}}_{c,\text{FB}}$ is strictly smaller than 1, where $\bar{\boldsymbol{M}}_{c,\text{FB}}$ is expressed as:

$$\bar{\boldsymbol{M}}_{c,\text{FB}} = \left(\bar{\boldsymbol{Z}}_{\text{Diag}} + \bar{\boldsymbol{Z}}_{\text{Forw}}\right)^{-1} \bar{\boldsymbol{Z}}_{\text{Forw}} \left(\bar{\boldsymbol{Z}}_{\text{Diag}} + \bar{\boldsymbol{Z}}_{\text{Back}}\right)^{-1} \bar{\boldsymbol{Z}}_{\text{Back}}. \quad [1.68]$$

This method will be tested in Chapter 2.

1.6. Random rough surface generation

In this section, we describe how to generate realizations of a random rough surface. We assume that the surface is Gaussian, which means that the height probability density function (PDF) follows a Gaussian process or a normal law. We assume that the surface height profile $z(x)$ is univocal and follows a stationary Gaussian random process.

1.6.1. *Statistical parameters*

For a centered Gaussian process, the surface height PDF is given by:

$$p_z(z) = \frac{1}{\sigma_z\sqrt{2\pi}} \exp\left(-\frac{z^2}{2\sigma_z^2}\right), \quad [1.69]$$

and checks:

$$\begin{cases} \langle 1 \rangle = \int_{-\infty}^{\infty} p_z(z)dz = 1 \\ \langle z \rangle = \int_{-\infty}^{\infty} zp_z(z)dz = 0 \\ \langle (z - \langle z \rangle)^2 \rangle = \langle z^2 \rangle = \int_{-\infty}^{\infty} z^2 p_z(z)dz = \sigma_z^2 \end{cases}, \quad [1.70]$$

where

$$\langle \bullet \rangle = \int_{-\infty}^{\infty} (\bullet) p_z(z)dz. \quad [1.71]$$

The real number σ_z stands for the surface height standard deviation and the surface height mean value $\langle z \rangle$ is zero. Since the height PDF is Gaussian, the derivative $d^n z(x)/dx^n$ also follows a Gaussian process.

Full characterization of the random rough surface height z is necessary for knowing the correlation between two heights on the surface of abscissa x_1 and x_2. For z real, the surface height autocorrelation function is then defined as:

$$\langle z(x_1) z(x_1 + x) \rangle = C_z(x). \quad [1.72]$$

Since the process is stationary, C_z depends only on the abscissa difference $x = x_2 - x_1$ between two points of the surface. Then, a Gaussian process is fully characterized by its height PDF, $p_z(z)$, and its surface height autocorrelation function, $C_z(x)$.

The power spectral density (PSD) or the surface height spectrum is defined as:

$$\hat{C}_z(k) = \text{FT}[C_z(x)] = \int_{-\infty}^{+\infty} C_z(x) e^{-jkx} dx, \quad [1.73]$$

and

$$C_z(x) = \frac{1}{2\pi} \text{FT}^{-1}[\hat{C}_z(k)] = \frac{1}{2\pi} \int_{-\infty}^{+\infty} \hat{C}_z(k) e^{jkx} dk, \quad [1.74]$$

where FT denotes the Fourier transform.

From equations [1.74], [1.73], [1.72] and [1.70], we have:

$$\sigma_z^2 = C_z(0) = \frac{1}{2\pi}\int_{-\infty}^{+\infty} \hat{C}_z(k)dk. \quad [1.75]$$

In addition, we can show that the surface slope autocorrelation function C_s is defined from the surface height autocorrelation function C_z as [BOU 99]:

$$C_s(x) = -\frac{d^2C_z}{dx^2} = \frac{1}{2\pi}\int_{-\infty}^{+\infty} k^2\hat{C}_z(k)e^{jkx}dx, \quad [1.76]$$

and then the slope variance is:

$$\sigma_s^2 = C_s(0) = \frac{1}{2\pi}\int_{-\infty}^{+\infty} k^2\hat{C}_z(k)dk. \quad [1.77]$$

In addition, equation [1.77] shows that the surface slope spectrum is $\hat{C}_s(k) = k^2\hat{C}_z(k)$.

1.6.2. *Generation of a random profile*

At the input of a linear filter, if e is a stationary process (of second order) of PSD $\hat{C}_e$, then the output signal s of PSD $\hat{C}_s$ satisfies [KUN 91]:

$$\hat{C}_s = \left|\hat{C}_g\right|^2 \hat{C}_e, \quad [1.78]$$

where $\hat{C}_g = \text{FT}(g)$ is the PSD of g, where g is the impulse response of the filter. In addition, if $\hat{C}_g \in \mathbb{R}^+$, then:

$$\hat{C}_g = \sqrt{\frac{\hat{C}_s}{\hat{C}_e}}. \quad [1.79]$$

Since the system is assumed to be linear, we have [KUN 91]:

$$s = g * e = \text{FT}^{-1}\left[\text{FT}(g)\text{FT}(e)\right] = \text{FT}^{-1}\left[\sqrt{\frac{\hat{C}_s}{\hat{C}_e}}\,\text{FT}(e)\right], \quad [1.80]$$

where the symbol $*$ stands for the convolution product.

Since we want to generate a surface height Gaussian process $z = s$, a Gaussian white noise of unitary variance is applied at the input of the filter, which implies that $\hat{C}_e = 1 \ \forall k$. Then:

$$z = \text{FT}^{-1}\left[\sqrt{\hat{C}_z}\,\text{FT}(e)\right]. \qquad [1.81]$$

Numerically, the convolution product is calculated in the Fourier domain because the complexity of a fast Fourier transform (FFT) is $\mathcal{O}(N \log N)$ instead of $\mathcal{O}(N^2)$ if the convolution product is calculated from its definition:

$$z(i) = g(i) * e(i) = \frac{1}{N}\sum_{n=1}^{N} g(n)e(n-i), \qquad [1.82]$$

where N is the length of both g and e. Since the surface height z is real, from equation [1.81], the function inside the operator FT^{-1} must satisfy $f^*(-k) = f(k) \ \forall k$, with * the complex conjugate operator. As shown further, $\hat{C}_z$ is real and an even function of k. Thus, $\text{FT}(e) = \hat{e}$ is a complex Gaussian white noise, which must satisfy $\hat{e}(-k)^* = \hat{e}(k)$. For more details, see [TSA 00, Chapter 4].

For surface height Gaussian and exponential autocorrelation functions defined as:

$$\begin{cases} C_z(x) = \sigma_z^2 \exp\left(-\dfrac{x^2}{L_c^2}\right) \\ C_z(x) = \sigma_z^2 \exp\left(-\dfrac{|x|}{L_c}\right) \end{cases}, \qquad [1.83]$$

the surface height spectra (or PSD) are given from equation [1.73] by:

$$\begin{cases} \hat{C}_z(k) = \sigma_z^2 L_c \sqrt{\pi} \exp\left(-\dfrac{k^2 L_c^2}{4}\right) \\ \hat{C}_z(k) = \dfrac{2\sigma_z^2 L_c}{1 + k^2 L_c^2} \end{cases}. \qquad [1.84]$$

For an exponential autocorrelation function, its derivative is not defined for $x = 0$. This implies, from equation [1.77], that the surface slope variance is not defined (from $-C_z''(0)$). On the other hand, it can be estimated from the spectrum of the generated surface $\hat{C}_z(k)$, because $\hat{C}_z(k)$ has a limited band, of upper cutoff spatial frequency k_c. Indeed, to have $k_c \to \infty$, the sampling step of the surface (generates from an FFT algorithm) must tend to zero since $\Delta x = \pi/(2k_c)$.

From equations [1.84], [1.83], [1.77] and [1.75], we can show for a Gaussian autocorrelation function that:

$$\begin{cases} \sigma_{z,k_c} \approx \sigma_{z,\infty} \text{erf}\left(\frac{k_c L_c}{2}\right) \\ \sigma_{s,k_c} \approx \sigma_{s,\infty} \left[\text{erf}\left(\frac{k_c L_c}{2}\right) - \frac{k_c L_c}{\sqrt{\pi}} \exp\left(-\frac{k_c^2 L_c^2}{4}\right)\right], \quad \sigma_{s,\infty} = \frac{\sigma_{z,\infty}\sqrt{2}}{L_c}, \end{cases} \quad [1.85]$$

and for an exponential autocorrelation function that:

$$\begin{cases} \sigma_{z,k_c} = \sigma_{z,\infty} \sqrt{\dfrac{2 \arctan(k_c L_c)}{\pi}} \\ \sigma_{s,k_c} = \sigma_{s,\infty} \sqrt{1 - \dfrac{\arctan(k_c L_c)}{k_c L_c}}, \quad \sigma_{s,\infty} = \sigma_{z,\infty} \sqrt{\dfrac{2k_c}{\pi L_c}}. \end{cases} \quad [1.86]$$

For $k_c \to \infty$, $\arctan(k_c L_c) \to \pi/2$ and $\arctan(k_c L_c)/(k_c L_c) \to 0$. Then, $\sigma_{z,\infty} = \sigma_z$ and $\sigma_{s,\infty} = \sigma_s$. Compared to a Gaussian autocorrelation function, the decreasing is slower since the function $\arctan(u)/u$ decreases more slowly than the functions $ue^{-u^2/4}/\sqrt{\pi}$ and erf$(u/2)$.

For remote sensing applications, it is interesting to study the case of a sea surface. Under approximations [ELF 97], the sea surface can be modeled as a Gaussian process. For a 2D surface, the surface height autocorrelation function is defined from its spectrum as:

$$C_z(x, y) = \frac{1}{(2\pi)^2} \int_{-\infty}^{+\infty} \int_{-\infty}^{+\infty} \hat{C}_z(k_x, k_y) e^{jk_x x + jk_y y} dk_x dk_y. \quad [1.87]$$

In addition, an ocean-like 2D spectrum is defined in polar coordinates (k_ρ, ϕ) from $\hat{C}_z(k_x, k_y)dk_xdk_y = k_\rho\hat{C}_{z\rho}(\rho,\phi)dk_\rho d\phi = S(k_\rho)[1+f(\phi)]dk_\rho d\phi$, where $S(k_\rho)$ is the isotropic part of the sea spectrum and $f(\phi)$ is its anisotropic part. As $C_z(0,0) = \sigma_z^2$, we then have:

$$\begin{aligned}\sigma_z^2 &= \frac{1}{(2\pi)^2}\int_{-\infty}^{+\infty}\int_{-\infty}^{+\infty}\hat{C}_z(k_x, k_y)dk_xdk_y \\ &= \frac{1}{(2\pi)^2}\int_0^{2\pi}[1+f(\phi)]\int_0^{\infty}S(k_\rho)dk_\rho \\ &= \frac{1}{2\pi}\int_0^{\infty}S(k_\rho)dk_\rho, \end{aligned} \quad [1.88]$$

since f is a periodic function over ϕ.

To set up a correspondence between a 2D spectrum $S(k_\rho)$ and 1D spectrum $\hat{C}_z(k)$, the comparison of equation [1.88] with equation [1.74] leads to $\hat{C}_z(|k|) = S(k_\rho)/2$ since $\hat{C}_z$ is an even function of k. In the following, the spectrum of Elfouhaily *et al.* is used [ELF 97] for dealing with the case of a sea surface.

1.6.3. *Simulations*

In Figure 1.4, the surface height autocorrelation function is Gaussian of correlation length $L_c = 10$ and of variance $\sigma_z = 1$. a) The number of samples is $N = 8{,}192$, and b) the number of samples is $N = 1{,}6384$.

The top of Figure 1.4 shows that the surface heights (more precisely, erf$(3/\sqrt(2))$ =99.7% of heights) range from $-3\sigma_z$ to $+3\sigma_z$. As the correlation length decreases, the simulations show (not displayed here) that the surface is more irregular (the horizontal distance between two consecutive extrema decreases) because the surface slope variance increases. Theoretically, from equation [1.85], if L_c is divided by 2, then the slope variance $\sigma_{s,\infty}$ is multiplied by 2.

In the middle of Figure 1.4, the histograms show that z follows a Gaussian process and as the number of samples N increases, the histogram better matches the theoretical histogram (which is related to the height PDF by a normalization).

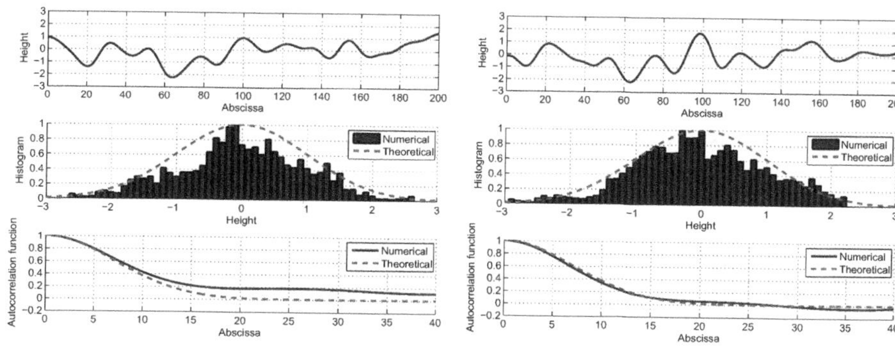

a) $N = 8{,}192$ b) $N = 16{,}384$

Figure 1.4. *Top: Surface heights over $x \in [0; 200]$ versus the abscissa x. Middle: Height histogram versus the surface heights. Bottom: Surface height autocorrelation function versus the abscissa x. The surface height autocorrelation function is Gaussian of correlation length $L_c = 10$ and of variance $\sigma_z = 1$. a) The number of samples is $N = 8{,}192$, and b) the number of samples is $N = 1{,}6384$*

The bottom of Figure 1.4 shows as N increases, the curve obtained numerically from z and equation [1.82] better matches the theoretical curve (equation [1.83]) for larger x. Indeed, as N increases, the length of the surface increases (with a constant sampling step) and then the correlation between far surface points is better predicted.

From Figure 1.5, the same remarks hold for an exponential autocorrelation function. The comparison of Figure 1.5 with Figure 1.4 shows that the surface is more irregular than that obtained from a Gaussian autocorrelation function, because the high frequencies contribute more to an exponential autocorrelation function. Indeed, as k increases, equation [1.84] shows that the corresponding spectrum decreases more slowly than that obtained from a Gaussian autocorrelation function. From an electromagnetic point of view, these high frequencies can have a strong impact on the scattered field.

Table 1.1 shows the values of the surface height and slope standard deviations, where σ_{z,k_c} and σ_{s,k_c} are computed from equations [1.85] and [1.86]. Moreover, $\tilde{\sigma}_z$ and $\tilde{\sigma}_s$ are computed from the generated surfaces shown in Figures 1.4 and 1.5. For the height variance, we can observe that σ_{z,k_c} is very close to $\tilde{\sigma}_z$ and the difference decreases as N increases. For the slope variance and for a Gaussian autocorrelation function, a good agreement is also obtained between σ_{s,k_c} and $\tilde{\sigma}_s$, whereas the agreement is less good for an exponential autocorrelation function. In addition, as N increases, the slope variance does not change significantly with a Gaussian autocorrelation function, whereas with an exponential autocorrelation function, it changes significantly.

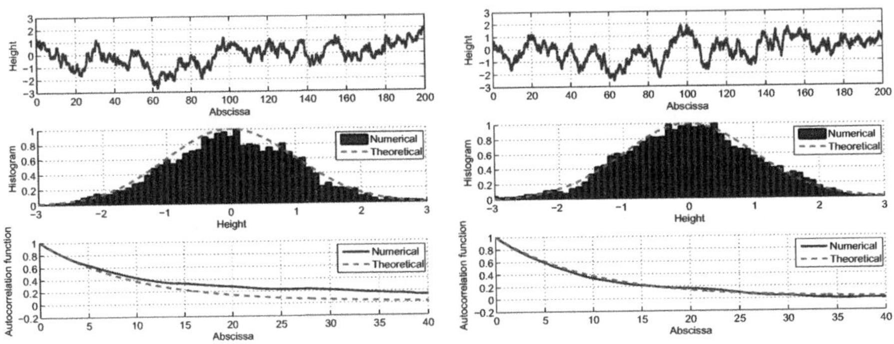

a) $N = 8{,}192$ b) $N = 16{,}384$

Figure 1.5. *Same plots as in Figure 1.4 but for an exponential autocorrelation function*

	N	$\tilde{\sigma}_z$	σ_{z,k_c}	$\tilde{\sigma}_s$	σ_{s,k_c}
Gaussian	8,192	0.963	1.000	0.132	0.141
Gaussian	16,384	0.988	1.000	0.141	0.141
Exponential	8,192	0.985	0.999	0.132	1.272
Exponential	16,384	0.993	0.999	1.587	1.805

Table 1.1. *Values of the surface height and standard deviations*

The comparison of the surface heights and slopes computed from a sea-like spectrum and a Gaussian autocorrelation function of the same height and slope variances is shown in Figure 1.6. The wind speed defined at 10 m above the sea is $u_{10} = 5$ m/s, the number of samples is $N = 524{,}288$ and the surface length is 400 m. Figure 1.6a) shows that the surface heights are similar, whereas the surface slopes strongly differ. Indeed, as shown in Figure 1.7, the sea slope spectrum contributes significantly to high wave numbers whereas for a Gaussian autocorrelation function, its slope spectrum rapidly decreases as k increases.

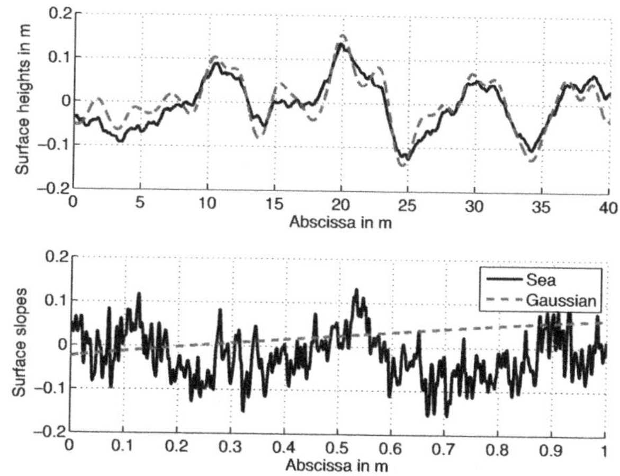

Figure 1.6. *Top: Surface heights in m versus the abscissa x in m. Bottom: Surface slopes versus the abscissa x in m. For the sea spectrum, the wind speed $u_{10} = 5$ m/s. The number of samples is $N = 524288$ and the surface length is 400 m. For the Gaussian surface height autocorrelation function, the height variance σ_z^2 is the same as the sea surface and its associated correlation length L_c is computed from the sea surface slope variance σ_s^2 and equation* [1.85] *with $k_c \to \infty$*

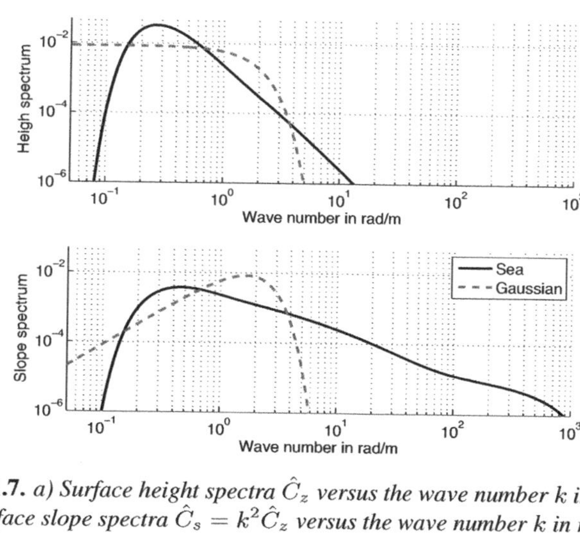

Figure 1.7. *a) Surface height spectra $\hat{C}_z$ versus the wave number k in rad/m. b) Surface slope spectra $\hat{C}_s = k^2\hat{C}_z$ versus the wave number k in rad/m. Same parameters as in Figure 1.6*

1.6.4. *Conclusion*

This section shows that the elevations of a random rough surface following a Gaussian process depend strongly on the choice of the height autocorrelation function. Indeed, two random rough surfaces having the same slope and height variances can strongly differ if the slope spectrum of one of the surfaces contributes to high frequencies. In fact, the surface variance (height, slope etc.) is obtained by integrating the corresponding spectrum over all the wave numbers, and cannot give the associated power for each frequency. When we want to properly characterize a natural surface (ocean, land, mountain, etc.), this point highlights that the knowledge of the surface height autocorrelation function or its spectrum is essential. Indeed, we will show in Chapter 2 that the high-frequency components of the surface can have a strong impact on the scattered field.

2

Validation of the Method of Moments for a Single Scatterer

2.1. Introduction

In this chapter, the method of moments (MoM), presented in Chapter 1, is tested on a cylinder of infinite length along the $\hat{y}$ direction. Indeed, from this canonical shape, the scattered field can be derived exactly and analytically by using the Bessel functions. Then, the scattered fields calculated with these two methods are compared for both a perfectly conducting and a dielectric cylinder. In addition, the MoM is compared with the physical optics (PO) approximation for the scattering from an elliptical cylinder, a plate and a rough surface.

Recall that the time convention $e^{-j\omega t}$ is used. Then, the derivative over the time t is $\partial/\partial t \to -j\omega$.

2.2. Solutions of a scattering problem

In this section, the possible ways to solve an electromagnetic scattering two-dimensional (2D) problem are described.

For a 2D problem and $\forall\ (\boldsymbol{r}, \boldsymbol{r}') \in S$, we have shown in Chapter 1 that the currents ψ_0 and $\partial\psi_0/\partial n$ on the surface S separating two dielectric media Ω_i ($i = \{0, 1\}$) satisfy:

$$\begin{cases} \psi_{\text{inc}}(\boldsymbol{r}') = -\displaystyle\int_S \left[\psi_0(\boldsymbol{r})\frac{\partial g_0(\boldsymbol{r}, \boldsymbol{r}')}{\partial n} - g_0(\boldsymbol{r}, \boldsymbol{r}')\frac{\partial\psi_0(\boldsymbol{r})}{\partial n}\right] dS \\ 0 \quad\quad\quad = \displaystyle\int_S \left[\psi_0(\boldsymbol{r})\frac{\partial g_1(\boldsymbol{r}, \boldsymbol{r}')}{\partial n} - \frac{1}{\rho_{01}}g_1(\boldsymbol{r}, \boldsymbol{r}')\frac{\partial\psi_0(\boldsymbol{r})}{\partial n}\right] dS \end{cases}, \qquad [2.1]$$

where $\boldsymbol{r}' = x'\hat{\boldsymbol{x}} + z'\hat{\boldsymbol{z}}$ and $\boldsymbol{r} = x\hat{\boldsymbol{x}} + z\hat{\boldsymbol{z}}$ are two points on the surface, and ψ_{inc} is the incident field in medium Ω_0 illuminating the surface. The problem is assumed to be invariant with respect to the $\hat{\boldsymbol{y}}$ direction. The Green function inside Ω_i is expressed as:

$$g_i(\boldsymbol{r}, \boldsymbol{r}') = \frac{j}{4}\mathrm{H}_0^{(1)}(k_i \left\|\boldsymbol{r} - \boldsymbol{r}'\right\|) = \frac{j}{4}\mathrm{H}_0^{(1)}\left[k_i\sqrt{(x - x')^2 + (z - z')^2}\right], \qquad [2.2]$$

where $\mathrm{H}_0^{(1)}$ is the zeroth-order Hankel function of the first kind. The scattered field $\psi_{\text{sca},i}$ inside Ω_i ($\boldsymbol{r}' \in \Omega_i$ and S excluded) is then computed using the Huygens' principle, in which the currents ψ_0 and $\partial\psi_0/\partial n$ on the surface are involved.

For the transverse electric (TE) polarization, the electric field $\boldsymbol{E}_i = \psi_i\hat{\boldsymbol{y}}$ ($i = \{0, 1\}$) and $\rho_{01} = 1$, whereas for the transverse magnetic (TM) polarization, the magnetic field $\boldsymbol{H}_i = \psi_i\hat{\boldsymbol{y}}$ and $\rho_{01} = \epsilon_0/\epsilon_1$, in which ϵ_i is the permittivity of Ω_i, which is assumed to be non-magnetic. The wave number $k_i = k_0 n_i = k_0\sqrt{\epsilon_{ri}}$, where ϵ_{ri} is the relative permittivity of the medium Ω_i and k_0 is the wave number in vacuum.

In addition, the field ψ_i satisfies the Helmholtz scalar wave equation in medium Ω_i:

$$\nabla^2\psi_i(\boldsymbol{r}) + k_i^2\psi_i(\boldsymbol{r}) = 0, \qquad [2.3]$$

and the boundary conditions

$$\begin{cases} \text{Dielectric media:} & \begin{cases} \psi_0(\boldsymbol{r}) = \psi_1(\boldsymbol{r}) \\ \hat{\boldsymbol{n}} \cdot \boldsymbol{\nabla}\psi_0(\boldsymbol{r}) = \rho_{01}\hat{\boldsymbol{n}} \cdot \boldsymbol{\nabla}\psi_1(\boldsymbol{r}) \end{cases} \\ \text{Dirichlet:} & \psi_0(\boldsymbol{r}) = 0 \\ \text{Neumann:} & \dfrac{\partial \psi_0(\boldsymbol{r})}{\partial n} = 0 \end{cases} . \qquad [2.4]$$

for $\boldsymbol{r} \in S$, and $\hat{\boldsymbol{n}}$ is the normal to the surface oriented toward Ω_0.

Finally, to have uniqueness of the solution, the scattered field must satisfy the far radiation condition (far from the object $r \to \infty$), which is expressed as [BOW 87]:

$$\lim_{r \to \infty} \sqrt{r} \left(\frac{\partial}{\partial r} - jk_i \right) \psi_i = 0. \qquad [2.5]$$

If the incident field is a plane wave, only the scattered field satisfies condition [2.5].

According to the scattering problem that needs to be solved, either the integral equations [2.1] or the Helmholtz equation [2.3] combined with the boundary conditions [2.4] and condition [2.5] are used.

For objects of simple shape, such as circular cylinder, elliptical cylinder and sphere for a three-dimensional problem, the scattered field can be derived with any approximation [BOW 87]. Section 2.3 derives the expression of the scattered field by an infinite cylinder illuminated by a plane wave.

For objects of complex shapes, the integral equations are solved numerically. One possible candidate is the MoM presented in Chapter 1. This method will be validated from the exact solution of the scattering from a circular cylinder and will be also used to study the accuracy of the PO approximation.

To solve the scattering problem, approximations can be introduced, allowing us to obtain an approximate expression (often analytical and applied next on a local region of the scatterer) of the scattered field. Section 2.4 presents the PO approximation.

2.3. Comparison with the exact solution of a circular cylinder in free space

In this section, the field scattered by a cylinder is discussed. Since the cylinder is assumed to be invariant with respect to the direction $\hat{\boldsymbol{y}}$, its length is infinite in this direction (see Figure 2.1). The **curl** operator in Cartesian ($\boldsymbol{F} = F_y(x, z)\hat{\boldsymbol{y}}$) and polar coordinates $(r,\theta\,)$ ($\boldsymbol{F} = F_r\hat{\boldsymbol{r}} + F_\theta\hat{\boldsymbol{\theta}}$) is defined as:

$$\textbf{curl}\, \boldsymbol{F} = -\frac{\partial F_y}{\partial z}\hat{\boldsymbol{x}} + \frac{\partial F_y}{\partial z}\hat{\boldsymbol{z}}, \qquad [2.6]$$

and

$$\textbf{curl}\, \boldsymbol{F} = \frac{1}{r}\frac{\partial F_y}{\partial \theta}\hat{\boldsymbol{r}} + \frac{\partial F_y}{\partial r}\hat{\boldsymbol{\theta}}. \qquad [2.7]$$

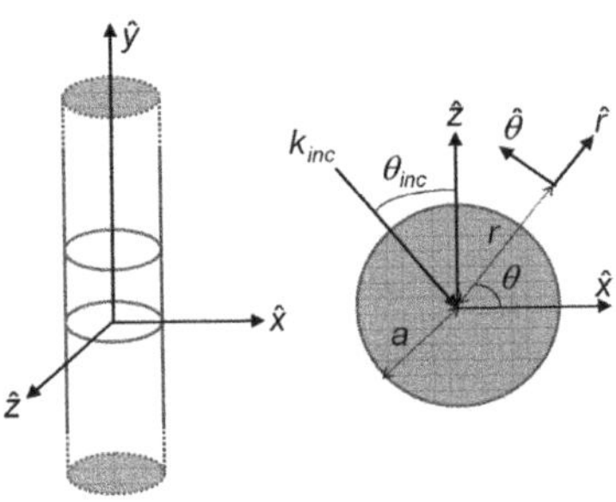

Figure 2.1. *Scattering from an infinite cylinder*

respectively. In addition

$$\boldsymbol{\nabla} = \frac{\partial}{\partial x}\hat{\boldsymbol{x}} + \frac{\partial}{\partial z}\hat{\boldsymbol{z}} = \frac{\partial}{\partial r}\hat{\boldsymbol{r}} + \frac{1}{r}\frac{\partial}{\partial \theta}\hat{\boldsymbol{\theta}}. \qquad [2.8]$$

With $\hat{\boldsymbol{n}} = \hat{\boldsymbol{r}}$, from equation [2.4] the boundary conditions [2.4] are then:

$$\begin{cases} \text{Dielectric media:} & \begin{cases} \psi_0(\boldsymbol{r}) = \psi_1(\boldsymbol{r}) \\ \dfrac{\partial \psi_0}{\partial r} = \rho_{01}\dfrac{\partial \psi_1}{\partial r} \end{cases} \\ \text{Dirichlet:} & \psi_0(\boldsymbol{r}) = 0 \\ \text{Neumann:} & \dfrac{\partial \psi_0(\boldsymbol{r})}{\partial r} = 0 \end{cases}. \qquad [2.9]$$

For the TE polarization, $\boldsymbol{E} = \psi(x,z)\hat{\boldsymbol{y}}$ and from a Maxwell equation [1.1], $\boldsymbol{H} = \textbf{curl}\,\boldsymbol{E}/(j\omega\mu_0)$. Then, from equations [2.6] and [2.7], we have:

$$\boldsymbol{H} = \frac{1}{j\omega\mu_0}\left(\frac{\partial\psi}{\partial z}\hat{\boldsymbol{x}} - \frac{\partial\psi}{\partial x}\hat{\boldsymbol{z}}\right) = \frac{1}{j\omega\mu_0}\left(\frac{1}{r}\frac{\partial\psi}{\partial\theta}\hat{\boldsymbol{r}} - \frac{\partial\psi}{\partial r}\hat{\boldsymbol{\theta}}\right). \quad [2.10]$$

For the TM polarization, $\boldsymbol{H} = \psi(x,z)\hat{\boldsymbol{y}}$ and from a Maxwell equation [1.1], $\boldsymbol{E} = -\textbf{curl}\,\boldsymbol{H}/(j\omega\epsilon)$. Then, from equations [2.6] and [2.7], we have:

$$\boldsymbol{E} = \frac{1}{j\omega\epsilon}\left(\frac{\partial\psi}{\partial x}\hat{\boldsymbol{z}} - \frac{\partial\psi}{\partial z}\hat{\boldsymbol{x}}\right) = \frac{1}{j\omega\epsilon}\left(\frac{\partial\psi}{\partial r}\hat{\boldsymbol{\theta}} - \frac{1}{r}\frac{\partial\psi}{\partial\theta}\hat{\boldsymbol{r}}\right). \quad [2.11]$$

2.3.1. *Solution of the Helmholtz equation*

The method of variable separation leads to $\psi_{\text{sca}}(r,\theta) = \psi_1(r)\psi_2(\theta)$. Since in polar coordinates:

$$\nabla^2 f = \frac{1}{r}\frac{\partial}{\partial r}\left(r\frac{\partial f}{\partial r}\right) + \frac{1}{r^2}\frac{\partial^2 f}{\partial\theta^2}, \quad [2.12]$$

the substitution of $\psi_{\text{sca}}(r,\theta)$ in the Helmholtz equation leads to:

$$\begin{aligned}
&\frac{1}{r}\frac{\partial}{\partial r}\left\{r\frac{\partial}{\partial r}\left[\psi_1(r)\psi_2(\theta)\right]\right\} + \frac{1}{r^2}\frac{\partial^2\left[\psi_1(r)\psi_2(\theta)\right]}{\partial\theta^2} + k^2\psi_1(r)\psi_2(\theta) = 0\\
&\Rightarrow \frac{\psi_2}{r}\left\{\frac{d\psi_1}{dr} + r\frac{d^2\psi_1}{dr^2}\right\} + \frac{\psi_1}{r^2}\frac{d^2\psi_2}{d\theta^2} + k^2\psi_1\psi_2 = 0\\
&\Rightarrow \frac{r^2}{\psi_1}\psi_1'' + \frac{r}{\psi_1}\psi_1' + k^2r^2 = -\frac{\psi_2''}{\psi_2}. \quad [2.13]
\end{aligned}$$

The left-hand side of the equality depends only on the variable r, whereas the right-hand side of the equality depends only on the variable θ. Thus, these two terms are equal if they are equal to the same constant $-\alpha^2$. This leads to:

$$\begin{cases}
\dfrac{\psi_2''(\theta)}{\psi_2(\theta)} = -\alpha^2\\
\dfrac{r^2}{\psi_1(r)}\psi_1''(r) + \dfrac{r}{\psi_1(r)}\psi_1'(r) + K^2r^2 = +\alpha^2
\end{cases}. \quad [2.14]$$

The solution of the first differential equation is $\psi_2(\theta) = e^{j\alpha\theta}$ (circular symmetry). In addition, since $\psi_2(\theta + 2\pi) = \psi_2(\theta)$, the constant $\alpha = n$ is an integer. The second differential equation is then:

$$r^2\psi_1''(r) + r\psi_1'(r) + (K^2r^2 - n^2)\psi_1(r) = 0. \qquad [2.15]$$

The solution of this differential equation can be expressed by the following Bessel functions [ABR 70]:

$$\begin{cases} \mathrm{J}_n(kr) \text{ Bessel function of first kind and order } n \\ \mathrm{Y}_n(kr) \text{ or } N_n(kr) \text{ Neumann function of first kind and order } n \end{cases}, \qquad [2.16]$$

where $\mathrm{Y}_n(x) = [\mathrm{J}_n(x)\cos(n\pi) - \mathrm{J}_{-n}(x)]/\sin(n\pi)$ [ABR 70].

Moreover, all linear combinations of both of these solutions, as, for instance, the Hankel functions defined by [ABR 70]:

$$\begin{cases} \mathrm{H}_n^{(1)}(kr) = \mathrm{J}_n(Kr) + j\mathrm{Y}_n(kr) \text{ Hankel function of first kind and order } n \\ \mathrm{H}_n^{(2)}(kr) = \mathrm{J}_n(Kr) - j\mathrm{Y}_n(kr) \text{ Hankel function of second kind and order } n \end{cases}, \qquad [2.17]$$

are also solutions to the differential equation.

In conclusion, the solution of the scalar wave propagation equation in polar coordinates is:

$$\begin{aligned} \psi_{\text{sca}}(r,\theta) &= \sum_{n=-\infty}^{n=+\infty} [A_n\mathrm{J}_n(kr) + B_n\mathrm{Y}_n(kr)]\, e^{jn\theta} \\ &= \sum_{n=-\infty}^{n=+\infty} \left[C_n\mathrm{H}_n^{(1)}(kr) + D_n\mathrm{H}_n^{(2)}(kr)\right] e^{jn\theta}, \end{aligned} \qquad [2.18]$$

where $A_n = C_n + D_n$ and $B_n = i(C_n - D_n)$.

The series expansions of the functions $\mathrm{H}_n^{(1)}$ and $\mathrm{H}_n^{(2)}$ are expressed as [ABR 70]:

$$\begin{cases} \mathrm{H}_n^{(1)}(z) = \sqrt{\dfrac{2}{\pi z}} e^{j\left(z-\frac{1}{2}n\pi-\frac{1}{4}\pi\right)} \\ \mathrm{H}_n^{(2)}(z) = \sqrt{\dfrac{2}{\pi z}} e^{-j\left(z-\frac{1}{2}n\pi-\frac{1}{4}\pi\right)} \end{cases} \text{when } |z| \to \infty, \qquad [2.19]$$

leading for $|z| \to \infty$ to

$$\begin{cases} \dfrac{\partial \mathrm{H}_n^{(1)}(z)}{\partial z} - j\mathrm{H}_n^{(1)}(z) = -\dfrac{1}{\sqrt{2z\pi z}} e^{j\left(z-\frac{n\pi}{2}-\frac{\pi}{4}\right)} \cong \dfrac{1}{z\sqrt{z}} \\ \dfrac{\partial \mathrm{H}_n^{(2)}(z)}{\partial z} - j\mathrm{H}_n^{(2)}(z) = -\dfrac{2j\sqrt{2}}{\sqrt{z\pi}} e^{-j\left(z-\frac{n\pi}{2}-\frac{\pi}{4}\right)} \cong \dfrac{1}{\sqrt{z}} \end{cases}$$

Therefore, the function $\mathrm{H}_n^{(2)}(Kr)$ does not satisfy the far radiation condition [2.5] and cannot be a solution ($D_n = 0$). The solution is then simplified as:

$$\psi_{\text{sca}}(r,\theta) = \sum_{n=-\infty}^{n=+\infty} C_n \mathrm{H}_n^{(1)}(kr) e^{jn\theta}, \qquad [2.20]$$

where the constant C_n will be determined from the boundary conditions.

Figure 2.2 shows the real and imaginary parts of $\mathrm{H}_n^{(1)}(x) \in \mathbb{C}$ versus x and n and the Bessel functions $\mathrm{J}_n(x) \in \mathbb{R}$ versus x and n.

First, we consider the TE polarization and we assume a perfectly conducting cylinder; next the TM polarization (Neumann condition) and finally, a dielectric cylinder.

2.3.2. *Dirichlet boundary conditions*

For the Dirichlet boundary condition, from equation [2.9], on the cylinder of radius a, $\psi_0(r = a,\theta) = 0\ \forall\theta$, and the field inside Ω_1 is zero (that is inside the cylinder). This implies that $\psi_{\text{sca},0}(a,\theta) + \psi_{\text{inc}}(a,\theta) = 0\ \forall\theta$.

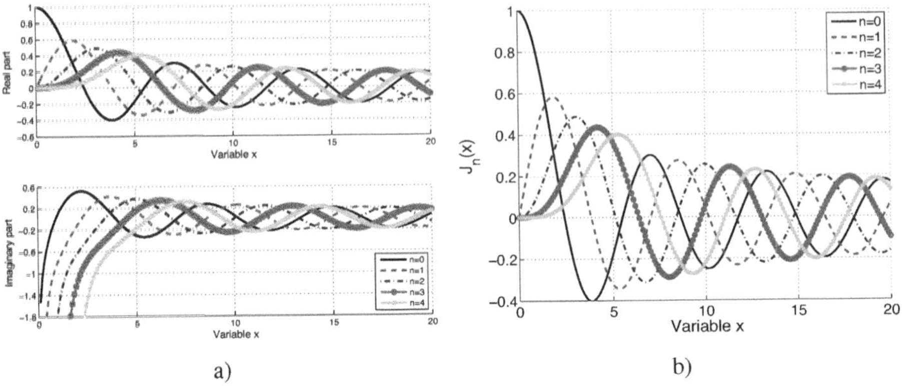

Figure 2.2. *a) Real and imaginary parts of* $H_n^{(1)}(x)$ *versus* x. *b) the Bessel functions* $\{J_n(x)\}$ *versus* x

Assuming a plane incident wave, $\psi_{\text{inc}} = \psi_{\text{inc},0} e^{jk_0(x \sin\theta_{\text{inc}} - z\cos\theta_{\text{inc}})} = \psi_{\text{inc},0} e^{jk_0 r \sin(\theta_{\text{inc}} - \theta)}$ ($\boldsymbol{k}_0 = k_0(\hat{\boldsymbol{x}} \sin\theta_{\text{inc}} - \hat{\boldsymbol{z}}\cos\theta_{\text{inc}})$) with $\tan\theta = z/x$ and $r = \sqrt{x^2+z^2}$, we have:

$$\psi_{\text{inc},0} e^{jk_0 a \sin(\theta_{\text{inc}} - \theta)} + \sum_{n=-\infty}^{n=+\infty} C_n \mathrm{H}_n^{(1)}(k_0 a) e^{-jn\theta} = 0 \; \forall \theta. \quad [2.21]$$

To have all the terms inside the sum, the plane wave is decomposed on the basis of the Bessel functions according to the [ABR 70]:

$$e^{\frac{1}{2}z\left(t - \frac{1}{t}\right)} = \sum_{n=-\infty}^{n=+\infty} t^n \mathrm{J}_n(z). \quad [2.22]$$

Setting $z = k_0 r$ and $t = e^{j(\theta_{\text{inc}} - \theta)}$, we have:

$$e^{jk_0 r \sin(\theta - \theta_{\text{inc}})} = \sum_{n=-\infty}^{n=+\infty} e^{jn(\theta_{\text{inc}} - \theta)} \mathrm{J}_n(k_0 r). \quad [2.23]$$

Then, equation [2.21] is:

$$\sum_{n=-\infty}^{n=+\infty} \left[\psi_{\text{inc},0} e^{jn\theta_{\text{inc}}} \mathrm{J}_n(k_0 a) + C_n \mathrm{H}_n^{(1)}(k_0 a)\right] e^{-jn\theta} = 0 \; \forall \theta. \quad [2.24]$$

This equality must be verified for any (θ, n), then:

$$C_n = -\frac{\psi_{\text{inc},0} e^{jn\theta_{\text{inc}}} \mathrm{J}_n(k_0 a)}{\mathrm{H}_n^{(1)}(k_0 a)}. \qquad [2.25]$$

In conclusion, for the TE polarization, the scattered field for a perfectly conducting cylinder is:

$$\psi_{\text{sca},0}(r,\theta) = -\psi_{\text{inc},0} \sum_{n=-\infty}^{n=+\infty} \frac{\mathrm{J}_n(k_0 a)}{\mathrm{H}_n^{(1)}(k_0 a)} \mathrm{H}_n^{(1)}(k_0 r) e^{jn(\theta_{\text{inc}}-\theta)}, \ r \geq a. \qquad [2.26]$$

To reduce the computing time, the following recurrence relation with $f_n(z) = \{\mathrm{J}_n(z), \mathrm{H}_n^{(1)}(z), \mathrm{H}_n^{(2)}(z)\}$ is used [ABR 70]:

$$f_{n+1}(z) = \frac{2n}{z} f_n(z) - f_{n-1}(z). \qquad [2.27]$$

Moreover, if n is an integer, then:

$$f_{-n}(z) = (-1)^n f_n(z). \qquad [2.28]$$

In conclusion, from $f_0(z)$ and $f_1(z)$, the sum for $n > 1$ can be computed and

$$\frac{\psi_{\text{sca},0}(r,\theta)}{\psi_{\text{inc},0}} = \sum_{n=1}^{n=+\infty} \frac{\mathrm{J}_n(k_0 a)}{\mathrm{H}_n^{(1)}(k_0 a)} \mathrm{H}_n^{(1)}(k_0 r) \left[e^{jn(\theta_{\text{inc}}-\theta)} + (-1)^n e^{-jn(\theta_{\text{inc}}-\theta)}\right] + \frac{\mathrm{J}_0(k_0 a)}{\mathrm{H}_0^{(1)}(k_0 a)} \mathrm{H}_0^{(1)}(k_0 r) \text{ for } r \geq a. \qquad [2.29]$$

2.3.3. *Neumann boundary conditions*

For the Neumann boundary condition, from equation [2.9], on the cylinder of radius a, $\partial\psi/\partial r = 0$. Thus, from equation [2.20], using the same process as in the TE polarization, we obtain with $\psi_{\text{inc}} = \psi_{\text{inc},0} \hat{\boldsymbol{y}} e^{jk_0 r \sin(\theta_{\text{inc}}-\theta)}$:

$$\psi_{\text{sca},0}(r,\theta) = \sum_{n=-\infty}^{n=+\infty} C_n \mathrm{H}_n^{(1)}(k_0 r) e^{-jn\theta}, \qquad [2.30]$$

where C_n satisfies the following boundary condition:

$$\frac{\partial}{\partial r}\left\{\sum_{n=-\infty}^{n=+\infty}\left[\psi_{\text{inc},0}e^{-jn\theta_{\text{inc}}}\text{J}_n(k_0r)+C_n\text{H}_n^{(1)}(k_0r)\right]e^{-jn\theta}\right\}\Bigg|_{r=a}=0\,\forall\theta.\text{[2.31]}$$

Then

$$C_n=-\frac{\psi_{\text{inc},0}e^{jn\theta_{\text{inc}}}\text{J}'_n(k_0a)}{\text{H}_n^{\prime(1)}(k_0a)},\qquad\text{[2.32]}$$

where the prime symbol denotes the derivative. The scattered field inside Ω_0 is then:

$$\psi_{\text{sca},0}(r,\theta)=-\psi_{\text{inc},0}\sum_{n=-\infty}^{n=+\infty}\frac{\text{J}'_n(k_0a)}{\text{H}_n^{\prime(1)}(k_0a)}\text{H}_n^{(1)}(Kr)e^{jn(\theta_{\text{inc}}-\theta)},\ r\geq a.\qquad\text{[2.33]}$$

Inside the cylinder, the scattered field $\psi_{\text{sca},1}$ is zero.

To avoid the need to numerically compute the deviations of the Bessel functions, the following recurrence relations with $f_n(x) = \{\text{J}_n(z),\text{H}_n^{(1)}(z),\text{H}_n^{(2)}(z)\}$ are applied [ABR 70]:

$$f'_n(z)=f_{n-1}(z)-\frac{n+1}{z}f_n(z)=\frac{n}{z}f_n(z)-f_{n+1}(z).\qquad\text{[2.34]}$$

Knowing f_0 and f_1, the derivatives f'_0 and f'_1 are calculated and from equations [2.27], [2.28] and [2.34], the sum over n can be reduced for $n > 0$.

Numerically, for the evaluation of equation [2.29] and [2.33], the sum over n must be truncated. Typically, the end value of n is chosen as the integer part of $2k_0a+1$ [BOW 87].

Figure 2.3a) shows the modulus of the total field versus the abscissa x and the height z for a perfectly conducting cylinder and for the TE polarization. Figure 2.3b) shows $|C_n|$ (equation [2.25]) versus the sum index n, $\lambda_0 = 1$ m, $a = 2\lambda_0$, $\psi_{\text{inc},0} = 1$ and $\theta_{\text{inc}} = 0$. Figure 2.4 shows the same results as in Figure 2.3, but for the TM polarization.

Figures 2.3a) and 2.4a) show behind (region $z < 0$) and near the cylinder that the total field is small, because the object is not directly illuminated on this side by the incident field. It corresponds to the so-called shadow zone. We can also observe the interference phenomenon between the scattered field and the incident field. As expected for the TE polarization, the total field is zero on the cylinder whereas for the TM polarization, the total field is maximized. Figures 2.3b) and 2.4b) show that the truncation of the sum at $n = N$, which equals the integer part of $2k_0a + 1$, is a good criterion since $|C_N| \approx 0$ and then, the higher order terms $n > N$ can be neglected in the calculation of the sum. For $a = 2\lambda_0$, $N = 27$.

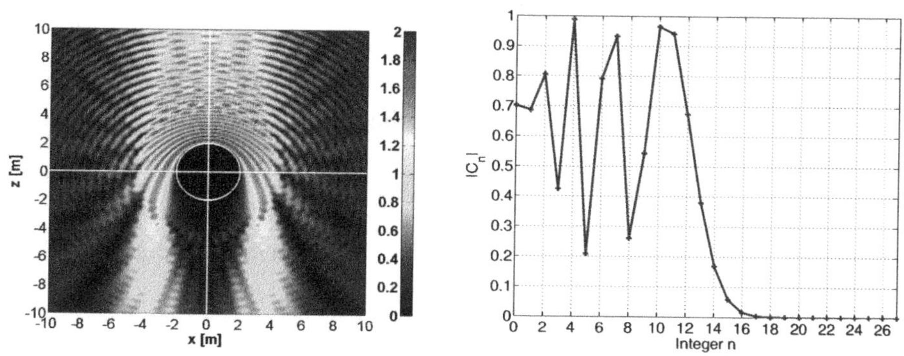

a) $|\psi_{\text{sca},0} + \psi_{\text{inc}}|$ b) $|C_n|$

Figure 2.3. *a) Modulus of the total field versus the abscissa x and the height z for a perfectly conducting cylinder and for the TE polarization. b) $|C_n|$ versus the sum index n. $\lambda_0 = 1$ m, $a = 2\lambda_0$, $\psi_{inc,0} = 1$ and $\theta_{inc} = 0$*

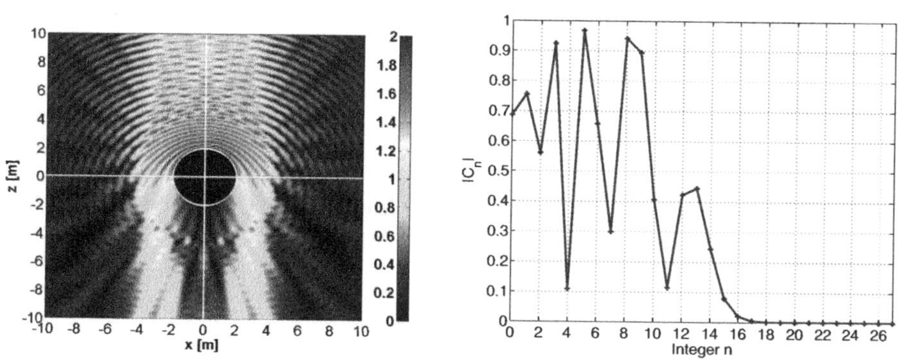

a) $|\psi_{\text{sca},0} + \psi_{\text{inc}}|$ b) $|C_n|$

Figure 2.4. *Same results as in Figure 2.3 but for the TM polarization*

2.3.4. *Dielectric cylinder*

For a dielectric cylinder (dielectric medium Ω_1), the Helmholtz scalar wave equation is:

$$\left[\nabla^2 + k_i^2(r)\right]\psi_i = 0, \tag{2.35}$$

where

$$k_i(r) = \begin{cases} k_0 \text{ for } r > a \\ k_1 \text{ for } r \leq a \end{cases}. \tag{2.36}$$

For $r > a$, from section 2.3.2, the total field inside Ω_0 can be written as (equations [2.20] and [2.23]):

$$\psi_0(r,\theta) = \sum_{n=-\infty}^{n=+\infty} \left[B_n \mathrm{J}_n(k_0 r) + C_n \mathrm{H}_n^{(1)}(k_0 r)\right] e^{-jn\theta} \text{ with } B_n = \psi_{\text{inc},0} e^{jn\theta_{\text{inc}}}. \tag{2.37}$$

For the field inside the cylinder ($r \leq a$), the first line of equation [2.18] is used, corresponding to a decomposition of an infinite sum of Bessel J_n and Neumann $\mathrm{Y}_n(x)$ functions. Moreover, since the Neumann function diverges for a null argument ($r = 0$, center of the cylinder), we have $B_n = 0$ in equation [2.18] and the field is then simplified as:

$$\psi_1(r,\theta) = \sum_{n=-\infty}^{n=+\infty} A_n \mathrm{J}_n(k_1 r) e^{-jn\theta}. \tag{2.38}$$

In equations [2.37] and [2.38], the two unknowns are C_n and A_n. From the boundary conditions [2.9], we have:

$$\begin{cases} \psi_0(a,\theta) = \psi_1(a,\theta) \\ \left.\dfrac{\partial\psi_0(r,\theta)}{\partial r}\right|_{r=a} = \rho_{01} \left.\dfrac{\partial\psi_1(r,\theta)}{\partial r}\right|_{r=a} \end{cases} \tag{2.39}$$

Thus,

$$\begin{cases} \sum_{n=-\infty}^{n=+\infty} \left[B_n \mathrm{J}_n(k_0 a) + C_n \mathrm{H}_n^{(1)}(k_0 a)\right] e^{-jn\theta} = \sum_{n=-\infty}^{n=+\infty} A_n \mathrm{J}_n(k_1 a) e^{-jn\theta} \\ k_0 \sum_{n=-\infty}^{n=+\infty} \left[B_n \mathrm{J}'_n(k_0 a) + C_n \mathrm{H}_n'^{(1)}(k_0 a)\right] e^{-jn\theta} = \rho_{01} k_1 \sum_{n=-\infty}^{n=+\infty} A_n \mathrm{J}'_n(k_1 a) e^{-jn\theta} \end{cases}$$

[2.40]

The above equation must be satisfied for all (θ, n). Then

$$\begin{cases} B_n \mathrm{J}_n(k_0 a) + C_n \mathrm{H}_n^{(1)}(k_0 a) = A_n \mathrm{J}_n(k_1 a) \\ B_n \mathrm{J}'_n(k_0 a) + C_n \mathrm{H}_n'^{(1)}(k_0 a) = \rho_{01} \frac{k_1}{k_0} A_n \mathrm{J}'_n(k_1 a) \end{cases}, \quad [2.41]$$

leading with $\alpha = \rho_{01} k_1 / k_0 = \rho_{01} \sqrt{\epsilon_1 / \epsilon_0}$ to

$$\begin{cases} A_n = B_n \frac{\mathrm{H}_n^{(1)}(k_0 a)\mathrm{J}'_n(k_0 a) - \mathrm{H}_n'^{(1)}(k_0 a)\mathrm{J}_n(k_0 a)}{\alpha \mathrm{H}_n^{(1)}(k_0 a)\mathrm{J}'_n(k_1 a) - \mathrm{H}_n'^{(1)}(k_0 a)\mathrm{J}_n(k_1 a)} \\ C_n = B_n \frac{\mathrm{J}_n(k_1 a)\mathrm{J}'_n(k_0 a) - \alpha \mathrm{J}_n(k_0 a)\mathrm{J}'_n(k_1 a)}{\alpha \mathrm{H}_n^{(1)}(k_0 a)\mathrm{J}'_n(k_1 a) - \mathrm{H}_n'^{(1)}(k_0 a)\mathrm{J}_n(k_1 a)} \end{cases}. \quad [2.42]$$

Substituting equation [2.42] into equations [2.37] and [2.38], the total field can be computed inside and outside the cylinder. Using the recurrence relations [2.27], [2.28] and [2.34], the sum over n can be reduced for $n > 0$ and the derivatives can be calculated analytically.

Figure 2.5a) shows the modulus of the total field versus the abscissa x and the height z for a dielectric cylinder and for the TE polarization. Figure 2.5b) shows $|A_n|$ and $|C_n|$ (equation [2.42]) versus the sum index n, $\lambda_0 = 1$ m, $a = 2\lambda_0$, $\psi_{\text{inc},0} = 1$, $\theta_{\text{inc}} = 0$ and $\epsilon_{r1} = 4 + 0.05j$. Figure 2.6 shows the same results as in Figure 2.5, but for the TM polarization.

Unlike a perfectly conducting cylinder, the scattered field inside the cylinder is not zero. Like the case of a perfectly conducting cylinder, Figures 2.5b) and 2.6b) show that the truncation of the sum at $n = N$, which equals the integer part of $2k_0a + 1$, is a good criterion since $|A_N| \approx 0$ and $|C_N| \approx 0$.

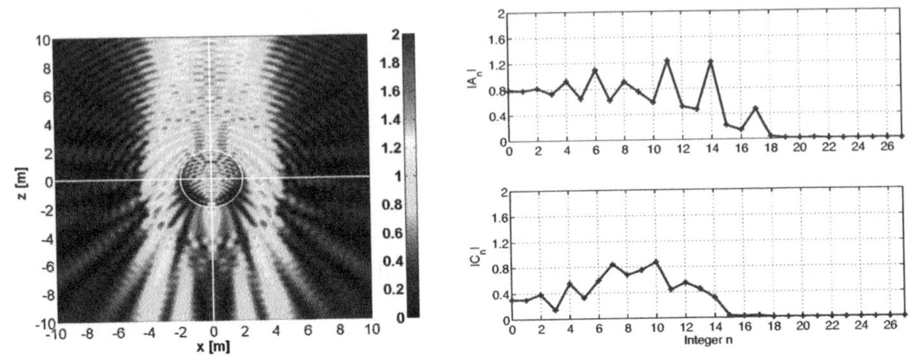

a) $|\psi_{\text{sca},0} + \psi_{\text{inc}}|$ b) $|A_n|$ and $|C_n|$

Figure 2.5. *a) Modulus of the total field versus the abscissa x and the height z for a dielectric cylinder and for the TE polarization. b) $|A_n|$ and $|C_n|$ versus the sum index n. $\lambda_0 = 1$ m, $a = 2\lambda_0$, $\psi_{inc,0} = 1$, $\theta_{inc} = 0$ and $\epsilon_{r1} = 4 + 0.05j$*

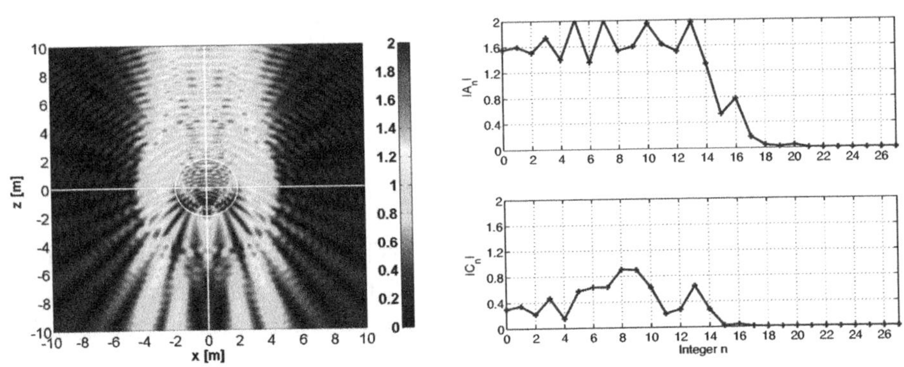

a) $|\psi_{\text{sca},0} + \psi_{\text{inc}}|$ b) $|A_n|$ and $|C_n|$

Figure 2.6. *Same results as in Figure 2.5, but for the TM polarization*

2.3.5. *MoM for an elliptical cylinder*

The MoM with point-matching method is thoroughly explained in section 1.3. The MoM converts the integral equation into a linear system as:

$$\bar{\boldsymbol{Z}}\boldsymbol{X} = \boldsymbol{b}, \quad [2.43]$$

where $\bar{\boldsymbol{Z}}$ is the impedance matrix, $\boldsymbol{X}$ is the unknown vector (currents on the sampled surface) and $\boldsymbol{b}$ is a vector of components equal to the incident field on the sampled surface.

For the Dirichlet boundary condition, the elements of $\bar{\boldsymbol{Z}}$ are expressed from equation [1.52], and the components of the vectors $\boldsymbol{X}$ and $\boldsymbol{b}$ are $X_n = \partial\psi_0(\boldsymbol{r}_n)/\partial n$ and $b_m = \psi_{\text{inc}}(\boldsymbol{r}_m)$, respectively. For the Neumann boundary condition, the elements of $\bar{\boldsymbol{Z}}$ are expressed by equation [1.55] and the components of the vectors $\boldsymbol{X}$ and $\boldsymbol{b}$ are $X_n = \psi_0(\boldsymbol{r}_n)$ and $b_m = \psi_{\text{inc}}(\boldsymbol{r}_m)$, respectively. For dielectric media, the elements of $\bar{\boldsymbol{Z}}$ are expressed by equation [1.58] and the vectors $\boldsymbol{X}$ and $\boldsymbol{b}$ are expressed by equation [1.59].

Using the Huygens' principle expressed in far field, the radar cross-section (RCS) is expressed as [1.31]:

$$\text{RCS} = \frac{\left|\psi_{\text{sca},0}^{\infty,0}\right|^2}{4\left|k_0\right|}, \quad [2.44]$$

where

$$\psi_{\text{sca},0}^{\infty,0} = -\frac{1}{\psi_{\text{inc},0}}\int_S \left[j\boldsymbol{k}_{\text{sca}}\cdot\hat{\boldsymbol{n}}\psi_0(\boldsymbol{r}) + \frac{\partial\psi_0(\boldsymbol{r})}{\partial n}\right] e^{-j\boldsymbol{k}_{\text{sca}}\cdot\boldsymbol{r}}dS. \quad [2.45]$$

In addition, $\psi_{\text{inc},0}$ is the modulus of the incident field in Ω_0 and the vector $\boldsymbol{k}_{\text{sca}}$ gives the observation direction.

The surface currents, ψ_0 and $\partial\psi_0/\partial n$, are computed from inverting the impedance matrix, and then the RCS is computed from the numerical integration on these surface currents via equation [2.45].

In polar coordinates, an elliptical cylinder of major and minor axis a and b is defined as:

$$\begin{cases} x = x_c + x_0 \cos\alpha - z_0 \sin\alpha \\ z = z_c + x_0 \sin\alpha + z_0 \cos\alpha \end{cases} \text{with} \begin{cases} x_0 = a\cos\theta \\ z_0 = b\sin\theta \end{cases}, \quad [2.46]$$

where (x_c, z_c) are the center coordinates of the cylinder and α the rotation angle defined from the direction $\hat{\boldsymbol{x}}$ and $\theta \in [0; 2\pi[$. Then, the slope on the cylinder surface is:

$$\gamma = \frac{dz}{dx} = \frac{dz_0 \cos\alpha + dx_0 \sin\alpha}{dx_0 \cos\alpha - dz_0 \sin\alpha} = \frac{\sin\alpha + \gamma_0 \cos\alpha}{\cos\alpha - \gamma_0 \sin\alpha}, \quad [2.47]$$

where

$$\gamma_0 = \frac{dz_0}{dx_0} = \frac{dz_0}{d\theta}\frac{d\theta}{dx_0} = \frac{b\cos\theta d\theta}{-a\sin\theta d\theta} = -\frac{b}{a}\cot\theta. \quad [2.48]$$

In addition, the derivative of γ is

$$\gamma' = \frac{d\gamma}{dx} = \frac{d\gamma}{d\theta}\frac{d\theta}{dx} = -\frac{ab}{(a\cos\alpha\sin\theta + b\sin\alpha\cos\theta)^3}. \quad [2.49]$$

The surface element $dS > 0$ is:

$$\begin{aligned} dS &= \sqrt{(dx)^2 + (dz)^2} = \sqrt{(dx_0)^2 + (dz_0)^2} = |dx_0|\sqrt{1+\gamma_0^2} \\ &= \sqrt{a^2 \sin^2\theta + b^2 \cos^2\theta} d\theta. \end{aligned} \quad [2.50]$$

The unitary vector normal to the cylinder $\hat{\boldsymbol{n}}$ is:

$$\hat{\boldsymbol{n}} = -\frac{\text{sgn}(dx)}{\sqrt{1+\gamma^2}}(-\gamma\hat{\boldsymbol{x}} + \hat{\boldsymbol{z}}). \quad [2.51]$$

For $\theta \in [0, \pi]$ (upper part of the cylinder), $-\text{sgn}(dx) = +1$, which is consistent with a rough surface, for which the normal to the surface always points toward Ω_0. For the Neumann boundary condition, the sampling step $\Delta_n = x_{n+1} - x_n$ must be substituted for $-|\Delta_n|\text{sgn}(dx)$ for the computation of the elements [1.55] of the impedance matrix, whereas for the Dirichlet boundary condition, the sampling step Δ_n must be substituted for $|\Delta_n|$ for the computation of the elements [1.52] of the impedance matrix.

To apply the MoM, the cylinder must be sampled as follows:

$$\theta_n = (n-1)\Delta\theta \text{ with } \Delta\theta = \frac{2\pi}{N} \text{ and } n \in [1; N], \quad [2.52]$$

where N_c is the number of samples describing the ellipse. We then define:

$$N_c = \frac{N_{\lambda_0} p}{\lambda_0}, \quad [2.53]$$

where N_{λ_0} is the number of samples per wavelength and p is the perimeter of the ellipse defined as:

$$p = \int_0^{2\pi} dS = 4\int_0^{\pi/2} \sqrt{a^2 \sin^2\theta + b^2\cos^2\theta d\theta}$$
$$= 4a\int_0^{\pi/2} \sqrt{1 - e^2\cos^2\theta d\theta} = 4a\int_0^1 \frac{\sqrt{1-e^2t^2}}{\sqrt{1-t^2}} dt, \quad [2.54]$$

where $e = \sqrt{|a^2 - b^2|}/\max(a, b)$ is the ellipse eccentricity. Then, p is computed from the "ellipke" Matlab function.

From θ_n, the coordinates (x_n, z_n), the slope γ_n, the sampling step Δ_n, $\text{sgn}(\Delta_n)$, and $\sqrt{1+\gamma_n^2} = \sqrt{a^2\sin^2\theta_n + b^2\cos^2\theta_n}$ of the ellipse are computed, which allows us to compute the elements of the impedance matrix. The incident wave is assumed to be a plane wave $\psi_{\text{inc}} = \psi_{\text{inc},0} e^{j\boldsymbol{k}_i \cdot \boldsymbol{r}} = \psi_{\text{inc},0} e^{jk_0 \sin(\theta_{\text{inc}} - \theta)}$, with $\boldsymbol{k}_i = k_0(\sin\theta_i \hat{\boldsymbol{x}} - \cos\theta_i \hat{\boldsymbol{z}})$.

2.3.6. *Numerical comparisons for a circular cylinder*

Here, the specific case of a circular cylinder of radius $a = b$, center $(x_x, z_c) = (0, 0)$ and rotation angle $\alpha = 0$ is considered, in order to validate the MoM with an exact analytical solution.

For the Dirichlet boundary condition, the total field on the cylinder is zero. To compare the exact solution with that obtained from the MoM, the normal derivative of $\psi_0(r,\theta) = \psi_{\text{sca},0}(r,\theta) + \psi_{\text{inc}}(r,\theta)$ must be derived. In polar coordinates, we have for a circular cylinder:

$$\frac{\partial\psi}{\partial n} = \nabla\psi\cdot\hat{\boldsymbol{n}} = \left(\frac{\partial\psi}{\partial r}\hat{\boldsymbol{r}} + \frac{1}{r}\frac{\partial\psi}{\partial\theta}\hat{\boldsymbol{\theta}}\right)\cdot\left(\frac{-\gamma\hat{\boldsymbol{x}}+\hat{\boldsymbol{z}}}{\sqrt{1+\gamma^2}}\right)\text{sgn}(\theta). \quad [2.55]$$

Knowing that $\hat{\boldsymbol{r}} = \cos\theta\hat{\boldsymbol{x}} + \sin\theta\hat{\boldsymbol{z}}$, $\hat{\boldsymbol{\theta}} = \sin\theta\hat{\boldsymbol{x}} - \cos\theta\hat{\boldsymbol{z}}$, $\gamma = -\cot\theta$, we have:

$$\begin{aligned}\frac{\partial\psi}{\partial n} &= \left[\frac{\partial\psi}{\partial r}(-\gamma\cos\theta+\sin\theta) + \frac{1}{r}\frac{\partial\psi}{\partial\theta}(-\gamma\sin\theta-\cos\theta)\right]\frac{1}{\sqrt{1+\gamma^2}}\\ &= \frac{\partial\psi}{\partial r}\text{sgn}^2(\sin\theta) = \frac{\partial\psi}{\partial r}.\end{aligned} \quad [2.56]$$

From equation [2.26], this leads, on the cylinder $r = a$, to:

$$\begin{aligned}\left.\frac{\partial\psi_0}{\partial r}\right|_{r=a} &= \left.\frac{\partial\psi_{\text{sca},0}}{\partial r}\right|_{r=a} + \left.\frac{\partial\psi_{\text{inc}}}{\partial r}\right|_{r=a}\\ &= k_0\psi_{\text{inc},0}\sum_{n=-\infty}^{n=+\infty}\left[\text{J}'_n(k_0a) - \frac{\text{J}_n(k_0a)}{\text{H}_n^{(1)}(k_0a)}\text{H}_n'^{(1)}(k_0a)\right]e^{jn(\theta_{\text{inc}}-\theta)}\\ &= \psi_{\text{inc},0}\sum_{n=-\infty}^{n=+\infty}\frac{2}{j\pi a\text{H}_n^{(1)}(k_0a)}e^{jn(\theta_{\text{inc}}-\theta)},\end{aligned} \quad [2.57]$$

since the Wonskrian $\text{H}_n^{(1)}(u)\text{J}'_n(u) - \text{H}_n'^{(1)}(u)\text{J}_n(u) = 2/(j\pi u)\ \forall(n,u)$ [ABR 70].

For the Neumann boundary condition, the field on the cylinder is expressed by equation [2.33] as:

$$\begin{aligned}\psi_0(a,\theta) &= \psi_{\text{inc},0}\sum_{n=-\infty}^{n=+\infty}\left[\text{J}_n(k_0a) - \frac{\text{J}'_n(k_0a)}{\text{H}_n'^{(1)}(k_0a)}\text{H}_n^{(1)}(k_0a)e^{jn(\theta_{\text{inc}}-\theta)}\right]\\ &= -\psi_{\text{inc},0}\sum_{n=-\infty}^{n=+\infty}\frac{2}{j\pi k_0a\text{H}_n'^{(1)}(k_0a)}e^{jn(\theta_{\text{inc}}-\theta)}.\end{aligned} \quad [2.58]$$

Using the same process, from equation [2.42] and [2.38], we can show that the currents on a dielectric cylinder are:

$$\begin{cases} \psi_0(a,\theta) = \psi_{\text{inc},0} \sum_{n=-\infty}^{n=+\infty} \dfrac{2\mathrm{J}_n(k_1 a)e^{jn(\theta_{\text{inc}}-\theta)}}{j\pi k_0 a \left[\alpha \mathrm{H}_n^{(1)}(k_0 a)\mathrm{J}'_n(k_1 a) - \mathrm{H}_n^{\prime(1)}(k_0 a)\mathrm{J}_n(k_1 a)\right]} \\ \left.\dfrac{\partial \psi_0}{\partial r}\right|_{r=a} = \psi_{\text{inc},0} \sum_{n=-\infty}^{n=+\infty} \dfrac{2\alpha \mathrm{J}'_n(k_1 a)e^{jn(\theta_{\text{inc}}-\theta)}}{j\pi a \left[\alpha \mathrm{H}_n^{(1)}(k_0 a)\mathrm{J}'_n(k_1 a) - \mathrm{H}_n^{\prime(1)}(k_0 a)\mathrm{J}_n(k_1 a)\right]} \end{cases} \quad [2.59]$$

For the Dirichlet boundary condition, $|\alpha| = |\sqrt{\epsilon_{r1}/\epsilon_{r0}}| \to \infty$, and then equation [2.57] is retrieved where $\psi_0(a,\theta) = 0$. For the Neumann boundary condition, $|\alpha| = |\sqrt{\epsilon_{r0}/\epsilon_{r1}}| \to 0$ and, then equation [2.58] is retrieved, where $\partial\psi_0(r,\theta)/\partial r|_{r=a} = 0$.

The RCS is expressed by equation [1.26]. Using equation [2.19] for the Taylor series expansion of the Hankel function when $r \to \infty$, from equations [2.26], [2.33], [2.42] and [2.38], and $\theta = \pi/2 - \theta_{\text{sca}}$ (from Figures 1.3 and 2.1), we have:

$$\text{RCS}(\theta_{\text{inc}}, \theta_{\text{sca}}) = \frac{4}{k_0} \left| \sum_{n=-\infty}^{n=+\infty} A_n e^{jn(\theta_{\text{inc}}+\theta_{\text{sca}}-\pi)} \right|^2, \quad [2.60]$$

where

$$A_n = \begin{cases} \dfrac{\mathrm{J}_n(k_0 a)}{\mathrm{H}_n^{(1)}(k_0 a)} & \text{Dirichlet} \\ \dfrac{\mathrm{J}'_n(k_0 a)}{\mathrm{H}_n^{\prime(1)}(k_0 a)} & \text{Neumann} \\ \dfrac{\mathrm{J}_n(k_1 a)\mathrm{J}'_n(k_0 a) - \alpha \mathrm{J}_n(k_0 a)\mathrm{J}'_n(k_1 a)}{\alpha \mathrm{H}_n^{(1)}(k_0 a)\mathrm{J}'_n(k_1 a) - \mathrm{H}_n^{\prime(1)}(k_0 a)\mathrm{J}_n(k_1 a)} & \text{Dielectric} \end{cases}, \quad [2.61]$$

and $\alpha = \sqrt{\epsilon_{r1}/\epsilon_{r0}}$ for the TE polarization and $\alpha = \sqrt{\epsilon_{r0}/\epsilon_{r1}}$ for the TM polarization.

In Figure 2.7, the label "MoM (N)" stands for the MoM, in which N is the number of unknowns on the cylinder, and the label "Ana (N_c)" stands for the exact analytical solution, in which N_c is the number of terms to compute the sum over n. Moreover, the RCS is plotted over the range $\theta_{\text{sca}} \in [0; \pi]$ because it is symmetric with respect to $\theta_{\text{inc}} = 0$.

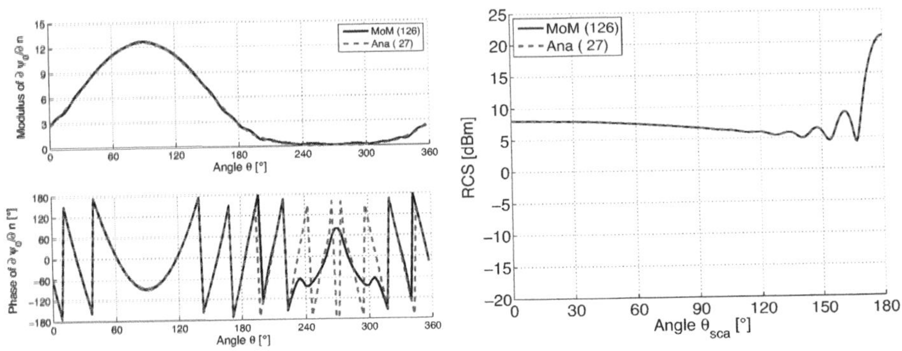

a) $|\partial\psi_0/\partial n|$ and $\arg(\partial\psi_0/\partial n)$

b) RCS

Figure 2.7. *a) Modulus* $|\partial\psi_0/\partial n|$ *and phase* $\arg(\partial\psi_0/\partial n)$ *on the cylinder for the Dirichlet boundary condition (TE polarization) versus the angle* θ. *b) RCS in dBm scale versus the observation angle* θ_{sca}. $\lambda_0 = 1$ *m,* $N_{\lambda_0} = 10$, $a = 2\lambda_0$ *and* $\theta_{inc} = 0$

Figure 2.7a) shows the modulus $|\partial\psi_0/\partial n|$ and the phase $\arg(\partial\psi_0/\partial n)$ on the cylinder versus the angle θ for the Dirichlet boundary condition (TE polarization). Figure 2.7b) shows the RCS in dBm scale versus the observation angle θ_{sca}. The simulations parameters are $\lambda_0 = 1$ m, $N_{\lambda_0} = 10$, $a = 2\lambda_0$ and $\theta_{\text{inc}} = 0$. Figure 2.8 shows the same results as in Figure 2.7 but for the Neumann boundary condition (TM polarization).

In Figure 2.7a), the modulus reaches its maximum in the direction $\theta = 90^\circ$ and decreases from this direction to reach its minimum behind the object defined for $\theta = 270^\circ$. This corresponds to the shadow zone. For the Dirichlet boundary condition and for $\theta = \pi/2$, $|\partial\psi_0/\partial n| = 2|\partial\psi_{\text{inc}}/\partial n| = 2|\partial\psi_{\text{inc}}/\partial r| = 2|jk_0 \sin(\theta_{\text{inc}} - \theta)\psi_{\text{inc}}| = 2k_0 \approx 12.57$ (also $\arg(\partial\psi_{\text{inc}}/\partial n) = -\pi/2$), whereas for the Neumann boundary condition, $|\psi_0| = 2|\psi_{\text{inc}}| = 2$ (also $\arg(\psi_0) = 0$). As shown, these levels can be predicted from the PO approximation. In addition, in Figure 2.8a), a perfect agreement between the two methods is observed, whereas the agreement is good in Figure 2.7a). In Figures 2.7b) and 2.8b), a perfect agreement is obtained between the two methods, and the RCS is maximum in the forward direction given by $\theta_{\text{sca}} = \pi + \theta_{\text{inc}}$. For this specific direction, it can be shown that the RCS is independent of the object shape and depends only on the silhouette of the object [KUB 11] in the high-frequency domain.

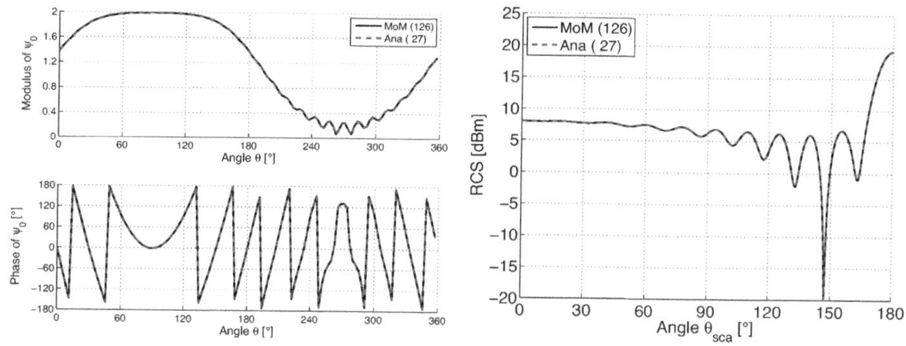

a) $|\psi_0|$ and $\arg(\psi_0)$ b) RCS

Figure 2.8. *Same results as in Figure 2.7 but for the Neumann boundary condition (TM polarization)*

Figure 2.9 shows the same results as in Figure 2.7 but for $N_{\lambda_0} = 20$. The increase of the number of samples slightly improves the agreement between the two methods.

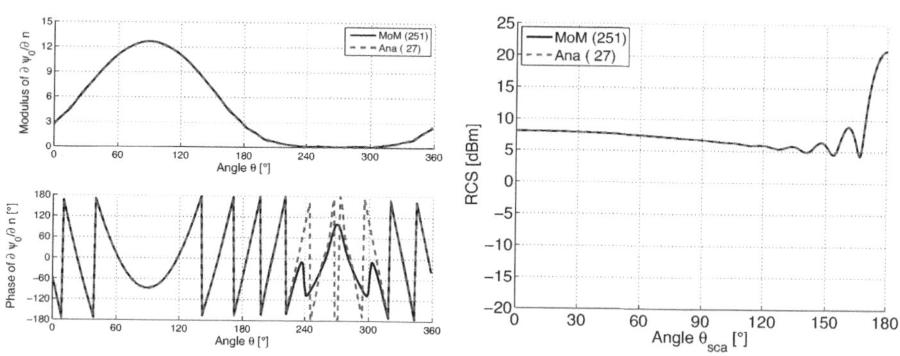

a) $|\partial\psi_0/\partial n|$ b) RCS

Figure 2.9. *Same results as in Figure 2.7 but for* $N_{\lambda_0} = 20$

Figure 2.10a) shows $|\psi_0|$, $\arg(\psi_0)$, $|\partial\psi_0/\partial r|$ and $\arg(\partial\psi_0/\partial r)$ versus the angle θ. The simulation parameters are $\lambda_0 = 1$ m, $a = 2\lambda_0$, $\psi_{\text{inc},0} = 1$, $\theta_{\text{inc}} = 0$ and $\epsilon_{r1} = 4 + 0.5j$ and the polarization is TE. Figure 2.10b) shows the same results as in Figure 2.10a), but for the TM polarization. Unlike a perfectly conducting cylinder (Figures 2.7a) and 2.8a)), the currents are maximum behind the object defined for $\theta = 270^{\circ}$, because a part of the incident field is transmitted into the cylinder. This phenomenon can also be observed in Figures 2.5 and 2.6. A perfect agreement between the two methods is observed.

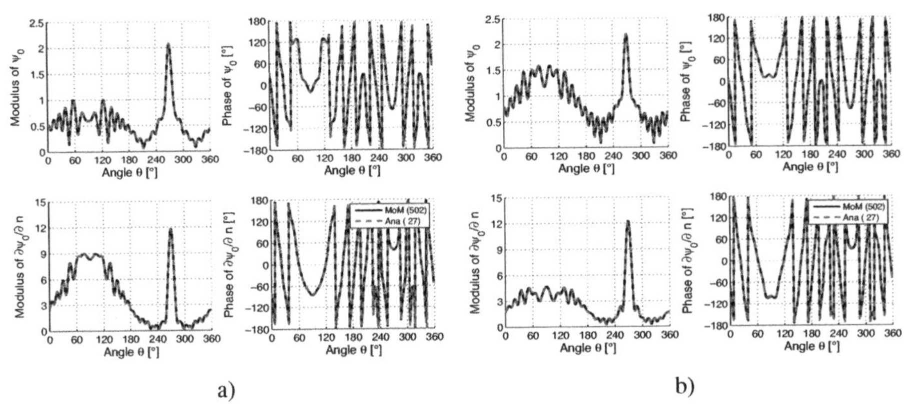

Figure 2.10. *Top:* $|\psi_0|$ *and* $\arg(\psi_0)$ *versus the angle* θ*. Bottom:* $|\partial\psi_0/\partial r|$ *and* $\arg(\partial\psi_0/\partial r)$ *versus the angle* θ*.* $\lambda_0 = 1$ *m,* $a = 2\lambda_0$*,* $\psi_{inc,0} = 1$*,* $\theta_{inc} = 0$ *and* $\epsilon_{r1} = 4 + 0.5j$*. a) TE polarization, b) TM polarization*

Figure 2.11 shows the RCS in dBm scale versus the angle θ_{sca} for the TE and TM polarizations. The simulation parameters are the same as in Figure 2.10a). Like a perfectly conducting cylinder (Figures 2.7b) and 2.8b)), the RCS is maximum in the forward direction defined as $\theta_{\text{sca}} = \pi + \theta_{\text{inc}} = \pi$ but its level is smaller than that obtained from a perfectly conducting cylinder.

For a dielectric cylinder, it is important to note that the number of samples per wavelength is $N_{\lambda_1} = N_{\lambda_0}|\sqrt{\epsilon_{r1}/\epsilon_{r0}}|$ because the wave number inside the cylinder is $k_1 = k_0\sqrt{\epsilon_{r1}}$. In addition, the number of unknowns is multiplied by two in comparison to a perfectly conducting cylinder, because both ψ_0 and $\partial\psi_0/\partial n$ must be computed.

Figure 2.12 shows the RCS in dBm scale, computed from the MoM, versus the angle θ_{sca} and for the TE polarization. $\lambda_0 = 1$ m, $a = 2\lambda_0$,

$\psi_{\text{inc},0} = 1$ and $\theta_{\text{inc}} = 0$. Figure 2.13 shows the same results as in Figure 2.12, but for the TM polarization. In Figure 2.12, the labels "MoM–PC (N)", "MoM–IBC (N)" and "MoM–DI (N)" stand for the MoM by considering perfectly conducting, IBC (highly conducting surface with impedance boundary condition) and dielectric surfaces, respectively. Moreover, N is the number of unknowns. For the IBC approximation, the elements of the impedance matrix are expressed by equation [1.62]. For both figures, the left figure a) shows results for $\epsilon_{r1} = 4 + 0.5j$, whereas the right figure b) shows results for $\epsilon_{r1} = 4 + 2j$.

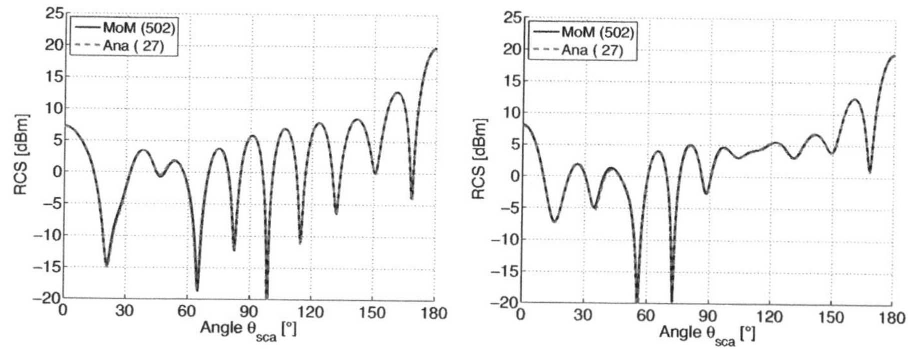

a) TE polarization

b) TM polarization

Figure 2.11. *RCS in dBm scale versus the angle θ_{sca}. Same simulation parameters as in Figure 2.10a)*

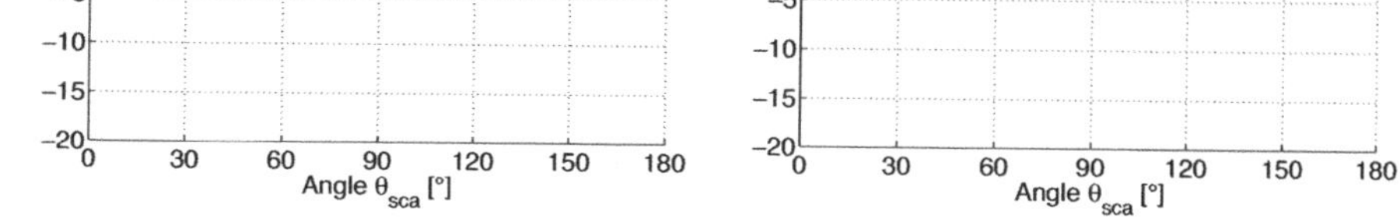

a) $\epsilon_{r1} = 4 + 0.5j$

b) $\epsilon_{r1} = 4 + 2j$

Figure 2.12. *RCS in dBm scale, computed from the MoM, versus the angle θ_{sca} and for the TE polarization. $\lambda_0 = 1$ m, $a = 2\lambda_0$, $\psi_{inc,0} = 1$ and $\theta_{inc} = 0$*

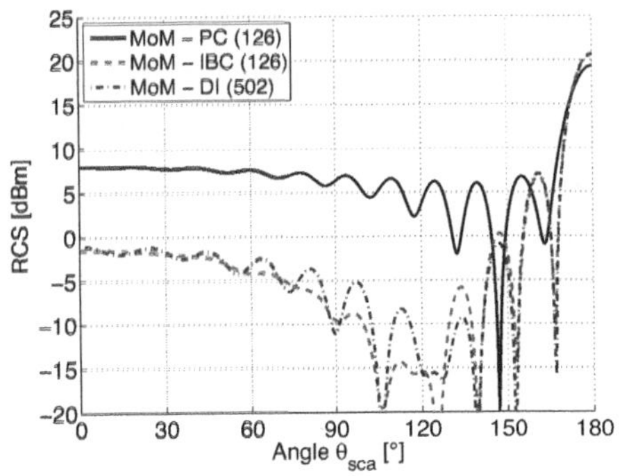

a) $\epsilon_{r1} = 4 + 0.5j$

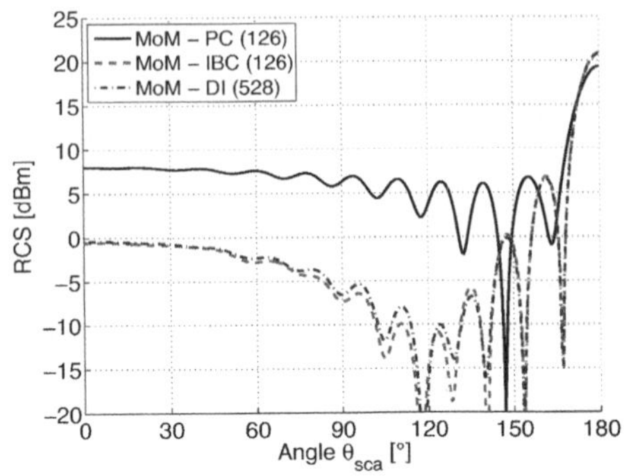

b) $\epsilon_{r1} = 4 + 2j$

Figure 2.13. *Same results as in Figure 2.12 but for the TM polarization*

As the imaginary part of the permittivity increases (from 0.5 to 2), the agreement between dielectric and IBC surfaces is better, especially for the TE polarization. The difference with a perfectly conducting surface remains significant.

The use of the IBC allows us to sample the surface along the wave number k_0 instead of k_1 and then the sampling step is independent of the permittivity of medium Ω_1. In terms of number of unknowns, it is equivalent to solve the scattering by a perfectly conducting surface, which makes the IBC very powerful.

2.3.7. *Conclusion*

By comparison with the exact solution, this section clearly shows that the MoM is efficient for calculating the field scattered by a (circular) cylinder. The advantage of the MoM is that it is able to numerically calculate the scattered field from an object of any shape. For example, for an elliptical cylinder, it is possible to obtain an exact solution [BOW 87] of the scattered field by introducing the Mathieu functions [ABR 70]. Nevertheless, unlike the Bessel functions, these functions are difficult to calculate numerically. It is, therefore, convenient to use the MoM for this specific shape.

As the frequency f increases ($k_0 = 2\pi/\lambda_0 = 2\pi f/c$ increases, where c is the celerity of light in a vacuum), the number of unknowns N increases and then the size ($N \times N$) of the impedance matrix increases. From a lower–upper (LU) inversion, the complexity (number of multiplications) is $\mathcal{O}(N^3)$ and then the inversion can require a long computing time. In addition, the impedance matrix must be stored to be inverted. The analytical solution requires the computation of a sum, where the number of terms is the integer part of $2k_0a + 1$. Then, at high frequencies, the series converges very slowly. In this case, the Watson transformation can be used to convert the solution into a rapidly convergent series. For more details, see [KON 05, Chapter 6].

It is then useful to develop methods to calculate the scattered field at high frequencies. It is discussed in the following section, which is focus on the PO approximation.

2.4. PO approximation

2.4.1. *Formulation*

For a surface illuminated by the incident field, the PO approximation assumes that a point on the surface $M(x, z)$ can be substituted for a straight line of infinite length which is tangent to the surface at this point. Then, the currents can be expressed by the Fresnel coefficient (valid for a planar surface of infinite area) and from a Snell–Descartes law giving the direction of the reflected field. In the shadowed zone S_{Sha}, the currents vanish. Then, the total field on the surface is:

$$\psi_0 = \begin{cases} [1 + \mathcal{R}(\theta)]\,\psi_{\text{inc}} & \boldsymbol{r} \in S_{\text{Ill}} \\ 0 & \boldsymbol{r} \in S_{\text{Sha}} \end{cases}, \qquad [2.62]$$

where $\mathcal{R}$ is the Fresnel coefficients and θ is the angle between the normal to the surface $\hat{\boldsymbol{n}}$ and the direction of the incident wave $\hat{\boldsymbol{k}}_{\text{inc}}$ (see Figure 2.14) defined as $\cos\theta = -\hat{\boldsymbol{k}}_{\text{inc}} \cdot \hat{\boldsymbol{n}}$. In addition, S_{Ill} stands for the illuminated surface and S_{Sha} stands for the shadowed surface ($S = S_{\text{Ill}} \cup S_{\text{Sha}}$). The PO approximation is valid for $\rho_0 \cos^3\theta \gg \lambda_0$ [BRE 80], where ρ_0 is the local radius of curvature of the surface.

Assuming a plane incident wave defined by $\psi_{\text{inc}} = \psi_{\text{inc},0} e^{j\boldsymbol{k}_{\text{inc}}\cdot\boldsymbol{r}}$, the use of the PO approximation leads to:

$$\begin{cases} \psi_0(\boldsymbol{r}) = [1 + \mathcal{R}(\theta(\boldsymbol{r}))]\,\psi_{\text{inc}}(\boldsymbol{r}) \\ \dfrac{\partial\psi_0(\boldsymbol{r})}{\partial n} = j\boldsymbol{k}_{\text{inc}} \cdot \hat{\boldsymbol{n}}(\boldsymbol{r})\,[1 - \mathcal{R}(\theta(\boldsymbol{r}))]\,\psi_{\text{inc}}(\boldsymbol{r}) \end{cases}, \quad \forall \boldsymbol{r} \in S_{\text{Ill}} \qquad [2.63]$$

and 0, otherwise. Substituting equation [2.63] into equation [2.45], the RCS under the PO approximation is:

$$\text{RCS} = \frac{\left|\psi_{\text{sca},0}^{\infty,0}\right|^2}{4\,|k_0|}, \tag{2.64}$$

where

$$\psi_{\text{sca},0}^{\infty,0} = -j\int_{S_{\text{III}}} \{\boldsymbol{k}_{\text{sca}}\cdot\hat{\boldsymbol{n}}\,[1+\mathcal{R}(\theta)] + \boldsymbol{k}_{\text{inc}}\cdot\hat{\boldsymbol{n}}\,[1-\mathcal{R}(\theta)]\}\, e^{j(\boldsymbol{k}_{\text{inc}}-\boldsymbol{k}_{\text{sca}})\cdot\boldsymbol{r}}dS. \tag{2.65}$$

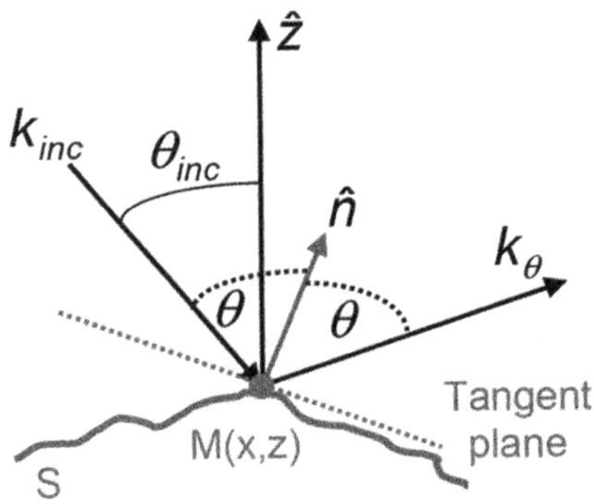

Figure 2.14. *Physical optics approximation*

2.4.2. *Applications*

2.4.2.1. *Perfectly conducting plate*

For a perfectly conducting plate (of length L and centered on (0,0), see Figure 2.15), the Fresnel reflection coefficient is $\mathcal{R}(\theta) = -1$ for the TE polarization and $\mathcal{R}(\theta) = +1$ for the TM polarization. In addition, the normal to the surface $\hat{\boldsymbol{n}} = \hat{\boldsymbol{z}}$ and $dS = dx$ for any $x \in S$. Equation [2.65] is then simplified as:

$$\psi_{\text{sca},0}^{\infty,0} = -2j\boldsymbol{k}_{\alpha}\cdot\hat{\boldsymbol{z}}\int_{-L/2}^{+L/2} e^{j(\boldsymbol{k}_{\text{inc}}-\boldsymbol{k}_{\text{sca}})\cdot\boldsymbol{x}}dx, \tag{2.66}$$

where $\boldsymbol{k}_\alpha = \boldsymbol{k}_{\text{inc}}$ for the TE polarization and $\boldsymbol{k}_\alpha = \boldsymbol{k}_{\text{sca}}$ for the TM polarization. The integration over x can then be done analytically starting from equation [2.64] to:

$$\text{RCS} = \frac{1}{k_0}\left\{k_0 L\cos\theta_\alpha \text{sinc}\left[\frac{k_0 L(\sin\theta_{\text{inc}} - \sin\theta_{\text{sca}})}{2}\right]\right\}^2, \qquad [2.67]$$

where $\text{sinc}(u) = \sin(u)/u$, $\theta_\alpha = \theta_{\text{inc}}$ for the TE polarization and $\theta_\alpha = \theta_{\text{sca}}$ for the TM polarization.

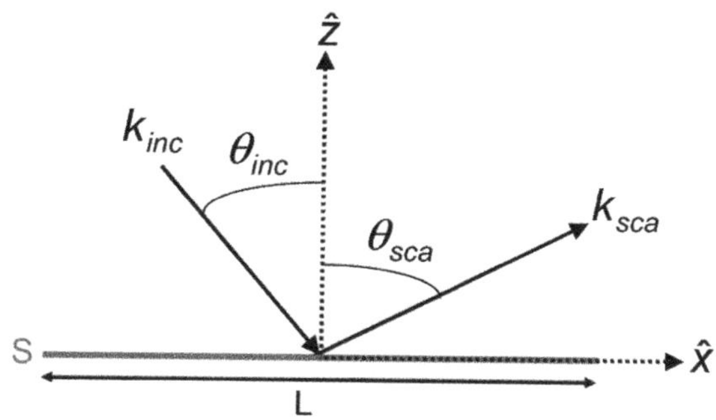

Figure 2.15. *Field scattered by a plate with the PO approximation*

In Figure 2.16, the label "MoM (N)" stands for the MoM, in which N is the number of unknowns on the plate, and the label "PO" stands for the PO solution. Moreover, the RCS is plotted over the range $\theta_{\text{sca}} \in [0; \pi]$ because it is symmetric with respect to $\theta_{\text{inc}} = 0$.

Figure 2.16a) shows the modulus $|\partial\psi_0/\partial n|$ and the phase $\arg(\partial\psi_0/\partial n)$ on the plate versus the integer $n \in [1; N]$ for the Dirichlet boundary condition. Figure 2.16b) shows the RCS in dBm scale versus the observation angle θ_{sca}, $\lambda_0 = 1$ m, $N_{\lambda_0} = 10$, $L = 5\lambda_0$, and $\theta_{\text{inc}} = 0$. Figure 2.17 shows the same results as in Figure 2.7, but for the Neumann boundary condition (TM polarization).

For the directly illuminated surface ($n \in [1; N/2]$), a good agreement is obtained between the PO approximation and the MoM. As expected, in the shadow region ($n \in [N/2 + 1; N]$), the PO predicts a null field, whereas the MoM predicts a small field, but does not equal zero due to the creeping waves. Moreover, the PO does not take into account the edge diffraction contribution coming from the fact that the surface has a finite length and explains why the field increases near the illuminated surface edges.

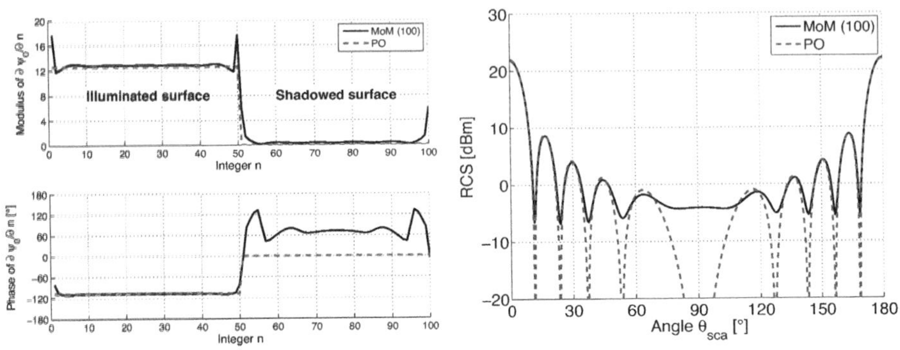

a) $|\partial\psi_0/\partial n|$ and $\arg(\partial\psi_0/\partial n)$ b) RCS

Figure 2.16. *a) Modulus $|\partial\psi_0/\partial n|$ and the phase $\arg(\partial\psi_0/\partial n)$ on the plate for the Dirichlet boundary condition versus the integer n. b) RCS in dBm scale versus the observation angle θ_{sca}. $\lambda_0 = 1$ m, $N_{\lambda_0} = 10$, $L = 5\lambda_0$ and $\theta_{inc} = 0$*

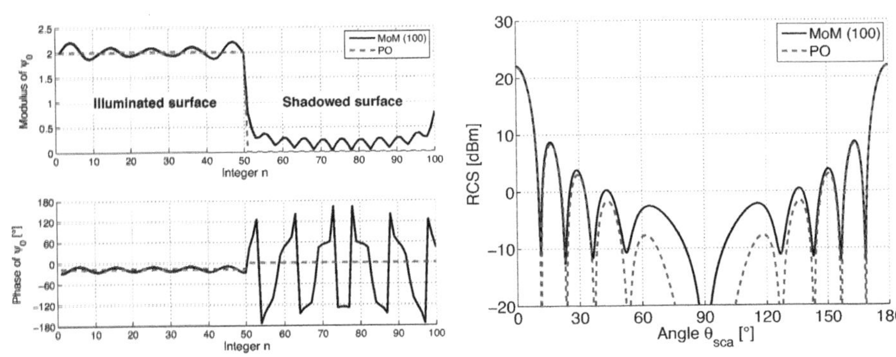

a) $|\psi_0|$ and $\arg(\psi_0)$ b) RCS

Figure 2.17. *Same results as in Figure 2.16, but for the TM polarization*

Figure 2.18 shows the same results as in Figure 2.16 but for $L = 10\lambda_0$. In comparison to Figure 2.16a), the agreement with the MoM is better because the surface length is twice longer, and then the currents related to the edge diffraction are more concentrated on the edges. On the RCS, the spreading around the specular ($\theta_{sca} = \theta_{inc} = 0$) and forward ($\theta_{sca} = \pi - \theta_{inc} = \pi$)

directions is smaller than that observed in Figure 2.16b) because the length L is twice longer. Like in Figures 2.16b) and 2.17b), far from these directions, the difference between the RCS computed from MoM and that from PO increases, but in this region, the levels are small.

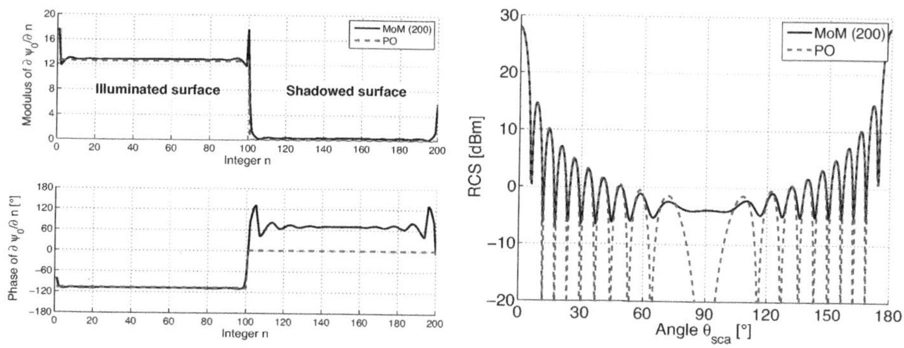

a) $|\partial\psi_0/\partial n|$ and $\arg(\partial\psi_0/\partial n)$

b) RCS

Figure 2.18. *Same results as in Figure 2.16 but* $L = 10\lambda_0$

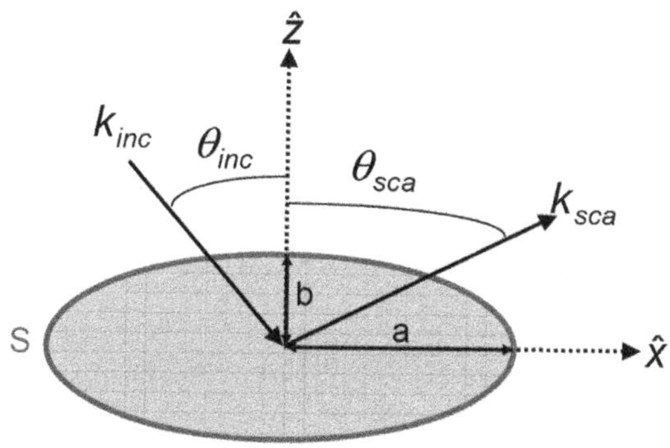

Figure 2.19. *Field scattered by a perfectly conducting elliptical cylinder with the PO approximation*

For the MoM, a plate thickness is introduced, which equals λ_0/N_{λ_0}. The advantage of introducing a thickness is to exhibit physical phenomena as the shadow and illuminated regions. The drawback is that the number of unknowns is twice as many. In fact, for the simulations, it is not necessary to introduce a thickness. Due to the linearity of the integral operator, the currents on the plate are then equal to the sum of the currents on the front and back sides (with thickness).

2.4.2.2. *Perfectly conducting elliptical cylinder*

In polar coordinates, the equation of an ellipse of center $(0, 0)$ is:

$$\begin{cases} x = a\cos\theta \\ z = b\sin\theta \end{cases}, \quad \theta \in [0; 2\pi[\qquad [2.68]$$

where a is the major axis and b is the minor axis. Moreover

$$\begin{aligned} dS\hat{\boldsymbol{n}} \cdot \boldsymbol{k}_{\text{sca}} &= \frac{-\gamma\hat{\boldsymbol{x}} + \hat{\boldsymbol{z}}}{\sqrt{1+\gamma^2}} \cdot \boldsymbol{k}_{\text{sca}}\sqrt{1+\gamma^2}\text{sgn}(\sin\theta)dx \\ &= (-\gamma\boldsymbol{k}_{\text{sca}} \cdot \hat{\boldsymbol{x}} + \boldsymbol{k}_{\text{sca}} \cdot \hat{\boldsymbol{z}})\,\text{sgn}(\sin\theta)dx, \end{aligned} \qquad [2.69]$$

where $\gamma = \frac{dz}{dx}$. Since $\gamma dx = dz = b\cos\theta d\theta$ and $dz = -a\sin\theta d\theta$, equation [2.63] is simplified as:

$$\begin{aligned} \psi_{\text{sca},0}^{\infty,0} = 2j \int_{S_{\text{III}}} & (b\cos\theta\boldsymbol{k}_\alpha \cdot \hat{\boldsymbol{x}} + a\sin\theta\boldsymbol{k}_\alpha \cdot \hat{\boldsymbol{z}}) \times \\ & e^{j[a(\boldsymbol{k}_{\text{inc}}-\boldsymbol{k}_{\text{sca}})\cdot\hat{\boldsymbol{x}}\cos\theta + b(\boldsymbol{k}_{\text{inc}}-\boldsymbol{k}_{\text{sca}})\cdot\hat{\boldsymbol{z}}\sin\theta]}\text{sgn}(\sin\theta)d\theta, \end{aligned} \qquad [2.70]$$

where $\alpha = \text{inc}$ for the TE polarization and $\alpha = \text{sca}$ for the TM polarization. Let $\cos\chi = \frac{a}{u}(\boldsymbol{k}_{\text{inc}} - \boldsymbol{k}_{\text{sca}}) \cdot \hat{\boldsymbol{x}}$ and $\sin\chi = \frac{b}{u}(\boldsymbol{k}_{\text{inc}} - \boldsymbol{k}_{\text{sca}}) \cdot \hat{\boldsymbol{z}}$, where:

$$\begin{cases} u = \sqrt{[a(\boldsymbol{k}_{\text{inc}} - \boldsymbol{k}_{\text{sca}}) \cdot \hat{\boldsymbol{x}}]^2 + [b(\boldsymbol{k}_{\text{inc}} - \boldsymbol{k}_{\text{sca}}) \cdot \hat{\boldsymbol{z}}]^2} \\ \tan\chi = \dfrac{b(\boldsymbol{k}_{\text{inc}} - \boldsymbol{k}_{\text{sca}}) \cdot \hat{\boldsymbol{z}}}{a(\boldsymbol{k}_{\text{inc}} - \boldsymbol{k}_{\text{sca}}) \cdot \hat{\boldsymbol{x}}} = -\dfrac{b}{a}\cot\left(\dfrac{\theta_{\text{inc}} - \theta_{\text{sca}}}{2}\right) \end{cases}, \qquad [2.71]$$

we obtain

$$\psi_{\text{sca},0}^{\infty,0} = 2j \int_{S_{\text{III}}} (b\cos\theta\boldsymbol{k}_\alpha \cdot \hat{\boldsymbol{x}} + a\sin\theta\boldsymbol{k}_\alpha \cdot \hat{\boldsymbol{z}})\, e^{ju\cos(\theta-\chi)}\text{sgn}(\sin\theta)d\theta. \qquad [2.72]$$

To easily calculate the illuminated zone, we assume that $\theta_i = 0$ and then the integration variable is $\theta \in [0; \pi]$. It is possible to treat the general case

$\theta_i \neq 0$ for which $\theta \in [\theta_{\min}; \theta_{\max}]$, in which $\theta_{\min}$ and $\theta_{\max}$ are calculated from the condition $\hat{\boldsymbol{n}} \cdot \hat{\boldsymbol{k}}_i = 0$ (limit between the illuminated and shadow regions).

From equation [2.22], we have:

$$e^{ju\cos(\theta-\chi)} = \sum_{n=-\infty}^{n=+\infty} j^n \mathrm{J}_n(u) e^{jn(\theta-\chi)}. \qquad [2.73]$$

Then

$$\begin{aligned}\int_0^{\pi} e^{ju\cos(\theta-\chi)} \cos\theta d\theta &= \sum_{n=-\infty}^{n=+\infty} j^n \mathrm{J}_n(u) e^{-jn\chi} \int_0^{\pi} e^{jn\theta} \cos\theta d\theta \\ &= \sum_{n=-\infty}^{n=+\infty} jnj^n \mathrm{J}_n(u) \frac{[1+(-1)^n]\, e^{-jn\chi}}{n^2-1}. \end{aligned} \qquad [2.74]$$

Using the same process, we show that:

$$\int_0^{\pi} e^{ju\cos(\theta-\chi)} \sin\theta d\theta = -\sum_{n=-\infty}^{n=+\infty} j^n \mathrm{J}_n(u) \frac{[1+(-1)^n]\, e^{-jn\chi}}{n^2-1}. \qquad [2.75]$$

Thus, the integration over $\theta \in [0; \pi]$ of equation [2.72] leads to:

$$\psi_{\text{sca},0}^{\infty,0} = 2j \sum_{n=-\infty}^{n=+\infty} j^n \mathrm{J}_n(u) \frac{[1+(-1)^n]\, e^{-jn\chi}}{n^2-1} \left(jnb\boldsymbol{k}_\alpha \cdot \hat{\boldsymbol{x}} - a\boldsymbol{k}_\alpha \cdot \hat{\boldsymbol{z}}\right). \qquad [2.76]$$

From equation [2.64], the RCS of an elliptical cylinder for $\theta_i = 0$ is:

$$\text{RCS} = \frac{1}{|k_0|} \left| \sum_{n=-\infty}^{n=+\infty} j^n \mathrm{J}_n(u) \frac{[1+(-1)^n]\, e^{-jn\chi}}{n^2-1} \left(jnb\boldsymbol{k}_\alpha \cdot \hat{\boldsymbol{x}} - a\boldsymbol{k}_\alpha \cdot \hat{\boldsymbol{z}}\right) \right|^2. \qquad [2.77]$$

For a circular cylinder, $a = b$, the above equation is valid $\forall \theta_i$ by making a rotation of the cylinder (due to the cylindrical symmetry). From equation [2.71], $\tan\chi = \cot([\theta_{\text{sca}} - \theta_{\text{inc}}]/2)$ and $u = 2a|\cos([\theta_{\text{sca}} - \theta_{\text{inc}}]/2)|$.

For a plate of length $L = 2a$, $b = 0$, $\sin\chi = 0$ and $\cos\chi = \text{sgn}[(\boldsymbol{k}_{\text{inc}} - \boldsymbol{k}_{\text{sca}})\cdot\hat{\boldsymbol{z}}] = \pm 1$. Then, $\chi = \pm\pi$, $e^{-jn\chi} = e^{\mp jn\pi} = (-1)^n$. In addition, $u = L|(\boldsymbol{k}_{\text{inc}} - \boldsymbol{k}_{\text{sca}})\cdot\hat{\boldsymbol{x}}|/2 = L|\sin\theta_{\text{inc}} - \sin\theta_{\text{sca}}|/2$. Thus

$$\text{RCS} = \frac{1}{k_0}(k_0 L\cos\theta_\alpha)^2\left|\frac{1}{2}\sum_{n=-\infty}^{n=+\infty} j^n \text{J}_n(u)\frac{(-1)^n+1}{n^2-1}\right|^2. \quad [2.78]$$

The comparison of equation [2.78] with equation [2.67] leads to:

$$\text{sinc}(u) = -\frac{1}{2}\sum_{n=-\infty}^{n=+\infty} j^n \text{J}_n(u)\frac{(-1)^n+1}{n^2-1}. \quad [2.79]$$

In Figure 2.20, the label "MoM (N)" stands for the MoM, in which N is the number of unknowns on the elliptical cylinder, and the label "PO" stands for the PO solution. Moreover, the RCS is plotted over the range $\theta_{\text{sca}} \in [0; \pi]$, because it is symmetric with respect to $\theta_{\text{inc}} = 0$.

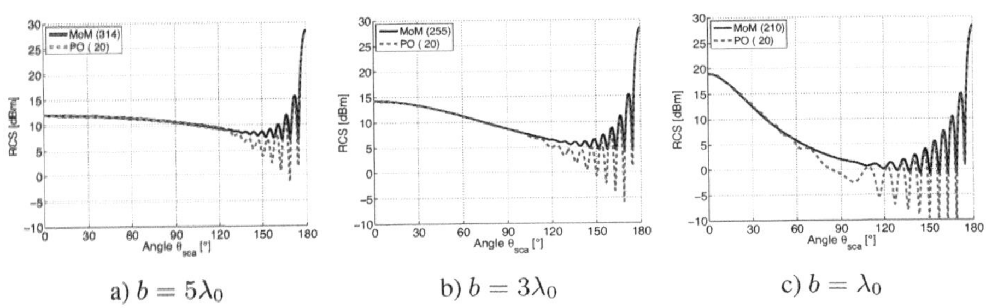

Figure 2.20. *RCS in dBm scale versus the observation angle θ_{sca}. $\lambda_0 = 1$ m, $N_{\lambda_0} = 10$, $a = 5\lambda_0$, $\theta_{inc} = 0$ and the polarization is TE*

Figure 2.20 shows the RCS in dBm scale versus the observation angle θ_{sca} and for different values of b. The simulation parameters are $\lambda_0 = 1$ m, $N_{\lambda_0} = 10$, $a = 5\lambda_0$, $\theta_{\text{inc}} = 0$ and the polarization is TE. Figure 2.21 shows the same results as in Figure 2.20 but for the TM polarization. Like for a plate, a good agreement between the MoM and PO is observed around the specular and the forward directions. Far from these directions, the difference increases between the two methods, but it remains smaller than that observed for a plate. Indeed, for an elliptical cylinder, the edge diffraction contribution is smaller. As b decreases, the region in which a difference is noticeable increases and in the forward direction, the RCS remains unchanged, whereas in the specular direction, the RCS increases. For more details on these phenomena, see [KUB 11].

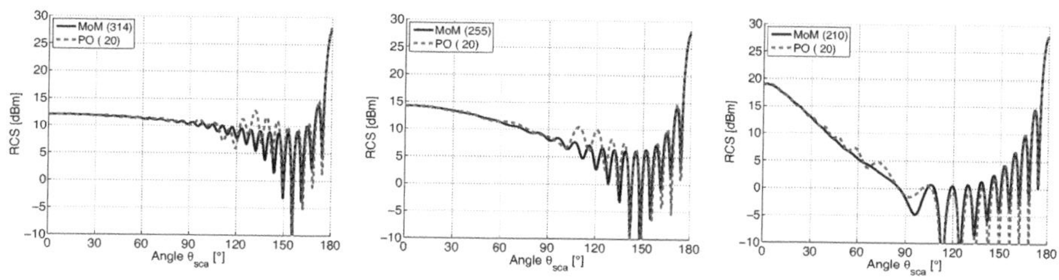

a) $b = 5\lambda_0$ b) $b = 3\lambda_0$ c) $b = \lambda_0$

Figure 2.21. *Same results as in Figure 2.20, but for the TM polarization*

2.4.2.3. *Gaussian random rough surface*

In this section, for a random rough surface, the PO is compared with the MoM. The generation of a rough surface is detailed in section 1.6. In addition, to avoid the edge diffraction phenomenon, the Thorsos tapered wave defined in section 1.2.5 is used as the incident field and the normalized radar cross-section (NRCS) is defined by equation [1.35]. For all simulations, the extent g is $g = L/6$ where L is the surface length.

Figure 2.22a) shows the surface heights versus the abscissa. Figure 2.22b) shows the associated NRCS in dB scale versus the observation angle θ_{sca} for the TE polarization and for a perfectly conducting surface. The surface height autocorrelation function is Gaussian, the wavelength is $\lambda_0 = 1$ m, the number of samples per wavelength is $N_{\lambda_0} = 10$, the surface length is $L = 300\lambda_0$, the height standard deviation is $\sigma_z = 0.5\lambda_0$, the correlation length is $L_c = 5\lambda_0$, the incidence angle is $\theta_{\text{inc}} = 30^{\text{o}}$ and the surface height autocorrelation function is Gaussian. Figure 2.23 shows the same results as in Figure 2.22b), but one parameter changes: (a) TM polarization, (b) $L_c = 2\lambda_0$, (c) $\sigma_z = \lambda_0$ and (d) $\epsilon_{r1} = 2 + 0.1j$.

Figure 2.22b) shows that the NRCS is maximum around the specular direction defined by $\theta_{\text{sca}} = \theta_{\text{inc}} = 30^{\text{o}}$ and that far from this direction, the NRCS decreases rapidly. For a smooth surface of infinite length, the NRCS is proportional to the Dirac delta function $\delta(\theta_{\text{sca}} - \theta_{\text{inc}})$ and then, the incident field is scattered only in the specular direction. As the correlation length decreases (divided by 2.5 in Figure 2.23b)) or σ_z increases (multiplied by 2 in Figure 2.23c)), the power is scattered in a larger angular domain θ_{sca} than that observed in Figure 2.22b).

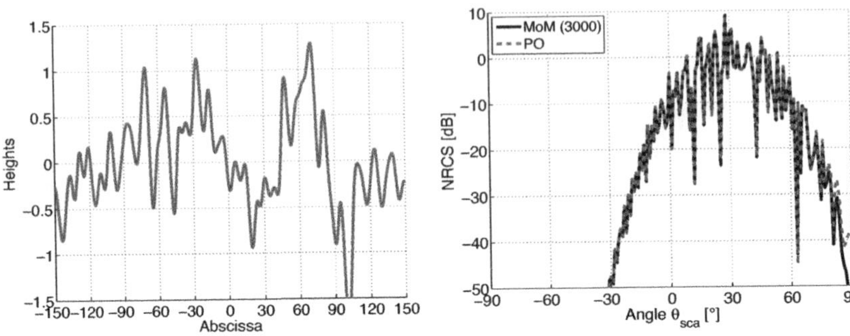

a) Surface heights z

b) NRCS

Figure 2.22. *a) Surface heights versus the abscissa. b) Associated NRCS in dB scale versus the observation angle θ_{sca} for the TE polarization and for a perfectly conducting surface. $\lambda_0 = 1$ m, $N_{\lambda_0} = 10$, $L = 300\lambda_0$, $\sigma_z = 0.5\lambda_0$, $L_c = 5\lambda_0$, $\theta_{inc} = 30^o$ and the surface height autocorrelation function is Gaussian*

In Figure 2.23b), the surface is more "irregular" because the slope standard deviation is larger (theoretically, $\sigma_s = \sqrt{2}\sigma_z/L_c$) than that of Figure 2.22a). This implies higher NRCS contributions far from the specular direction. In addition, for grazing angles ($|\theta_{\text{sca}}|$ approaching $\pi/2$), the results obtained from the two methods differ slightly because the PO (developed at the first order, here) does not take the multiple reflections into account. The PO can be extended to account for this phenomenon [ISH 86, BOU 04a, BOU 04b], but then the complexity increases. Besides, from the view point of the geometric optics approximation, due to the surface roughness, a part of the surface cannot be illuminated from the transmitter or/and the receiver. This phenomenon can also be integrated via a shadowing function [BOU 04a, BOU 04b, LI 11, PIN 13]. These two phenomena are illustrated in Figure 2.24.

In Figure 2.23c), the surface is rougher from the view point of the Rayleigh parameter defined as $R_a = k_0\sigma_z(\cos\theta_{\text{inc}} + \cos\theta_{\text{sca}})$ [OGI 91, PIN 13]. This parameter is related to the phase difference from a point of the surface to its mean plane. Thus, as σ_z increases, the surface becomes rougher and the power is more incoherent.

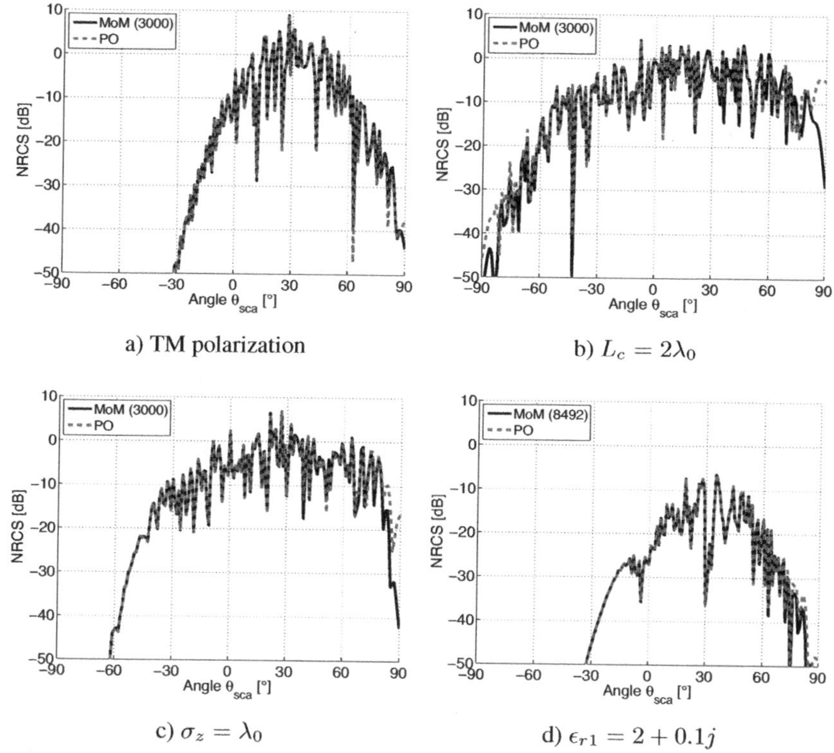

Figure 2.23. *Same results as in Figure 2.22b), but for a) TM polarization, b) $L_c = 2\lambda_0$, c) $\sigma_z = \lambda_0$ and d) $\epsilon_{r1} = 2 + 0.1j$*

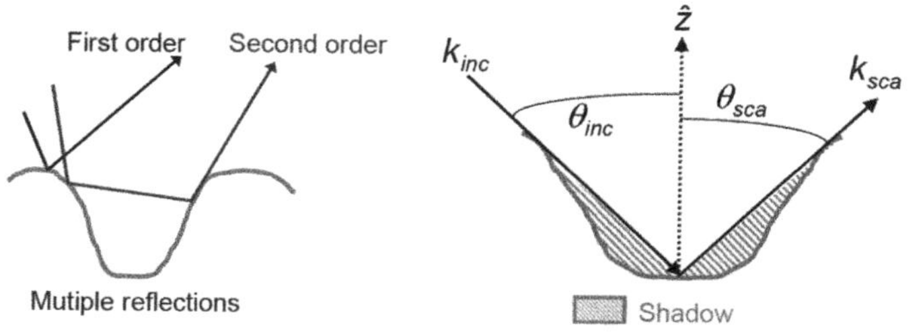

Figure 2.24. *Illustration of phenomena not accounted for in PO: multiple reflections and the shadow*

Figures 2.23a) and d) highlight a good agreement between the two methods, and for a lower dielectric medium, the NRCS levels are smaller than those obtained from a perfectly conducting surface, because a part of the scattered field is transmitted into the lower medium.

2.4.2.4. *Exponential random rough surface*

Figure 2.25 shows the same results as in Figure 2.22, but for an exponential surface height autocorrelation function. As expected, the surface elevations differ strongly from those of a Gaussian autocorrelation function (see Figure 2.22a)). As explained in section 1.6, this comes from the contribution of the surface high frequencies. The comparison of Figure 2.25b) with Figure 2.22b) clearly shows that two rough surfaces having the same correlation length and the same height standard deviation can give totally different NRCSs. In addition, Figure 2.25b) shows that the agreement between the PO and the MoM is poor because the PO can not correctly estimate the contribution of the surface high-frequencies. Indeed, for Figure 2.22a), the mean value of $\rho_0 \cos^3\theta$ is $106.77 \gg \lambda_0 = 1$, whereas for Figure 2.25a), $\langle \rho_0 \cos^3\theta \rangle \approx 0.66 < \lambda_0$.

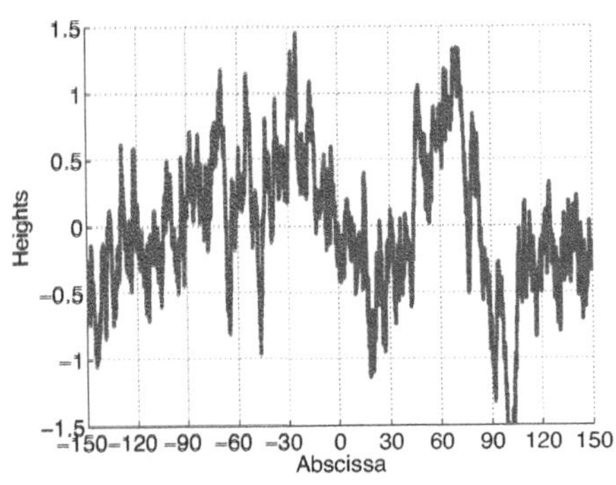

a) Surface heights z

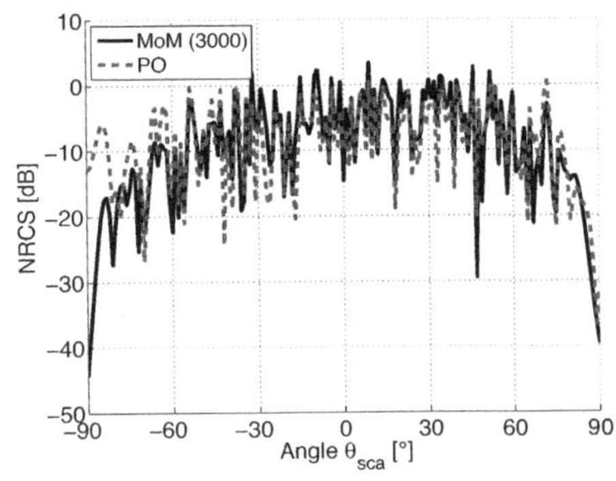

b) RCS

Figure 2.25. *Same results as in Figure 2.22, but for an exponential surface height autocorrelation function*

2.4.3. *Sea-like surface*

Figure 2.26 shows the same results as in Figure 2.22a), but for a sea surface. The wind speed, u_{10}, defined at 10 m above the sea is 5 m/s. The surface length is $L = 50$ m, the number of points is 5,000 and the surface height spectrum is that of Elfouhaily *et al.* [ELF 97]. Figure 2.27 shows the NRCS of this surface in dB scale versus the observation angle θ_{sca}: (a) TE polarization and (b) TM polarization. The simulation parameters are $\lambda_0 = 0.1$ m, $N_{\lambda_0} = 10$, $\epsilon_{r1} = 70.4 + 40.6j$ and $\theta_{\text{inc}} = 30^\circ$.

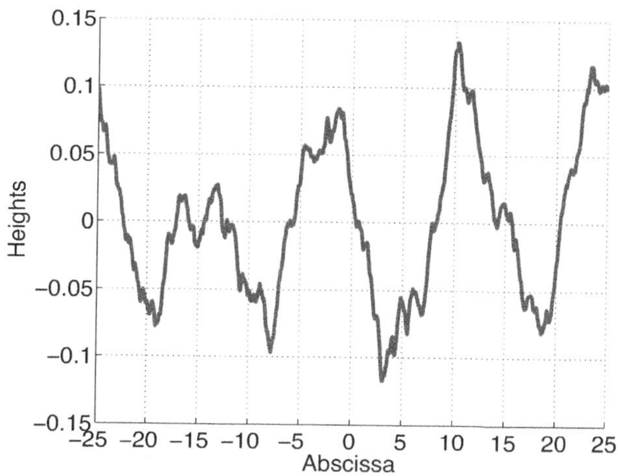

Figure 2.26. *Same results as in Figure 2.22a) but for a sea surface. The wind speed, u_{10}, defined at 10 m above the sea is 5 m/s. The surface length is $L = 50$ m, the number of points is 5,000 and the surface height spectrum is that of Elfouhaily et al. [ELF 97]*

Figure 2.26 shows that the surface length is large enough to observe several periods of the waves having the greatest wavelength. Since a fast Fourier transform (FFT) algorithm is applied to generate the surface, the sea spectrum is truncated at $k_{\text{min}} = \pi/L$, for the lower frequency, and at $k_{\text{max}} = \pi/\Delta x$, for the upper frequency. With a sampling step $\Delta x = \lambda_0/10$, we have $k_{\text{max}} = 10\pi/\lambda_0$. In [BOU 00], a criterion is given to ensure that all the frequencies of the surface height spectrum are generated: $k_{\text{min}} \leq 0.28 k_{\text{peak}}$ in which $k_{\text{peak}} = \Omega^2 g/u_{10}^2$ is the wave number of the most energetic wave, with $\Omega = 0.84$ for a fully developed sea and $g = 9.81$ m/s^2. Thus, with $u_{10} = 5$ m/s, $k_{\text{peak}} = 0.277$ rad/m, which implies that $k_{\text{min}} \leq 0.0775$ rad/m, and then $L \geq 40.5$ m. With $L < 40.5$ m, the low frequencies of the sea spectrum are truncated.

Figure 2.27 shows a good agreement between the MoM and PO methods and the agreement is better for the TM polarization. A better agreement can be obtained by using specific asymptotic models [PIN 13, ELF 04, BOU 05] developed for a sea surface. Since $|\epsilon_{r1}| \gg 1$, the IBC approximation is applied to calculate the scattered field. This allows us to decrease the number of unknowns if the surface is considered as dielectric. Otherwise, the number of unknowns should be approximately multiplied by $|\sqrt{\epsilon_{r1}}| \approx 9$ leading to 45,000.

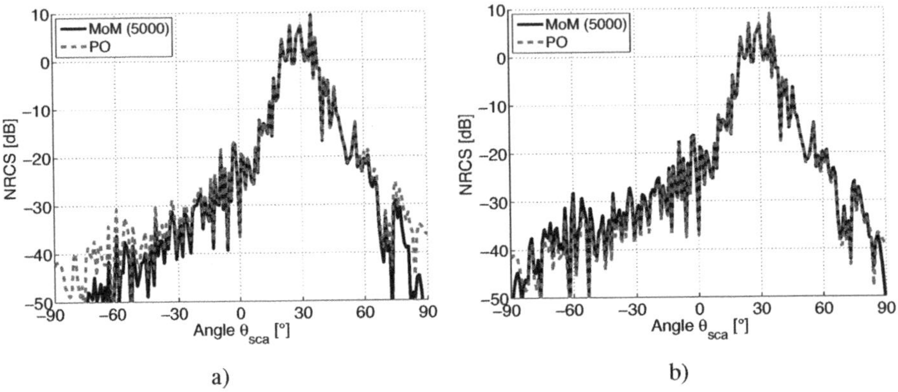

Figure 2.27. *NRCS in dB scale versus the observation angle θ_{sca}. a) TE polarization. b) TM polarization. $\lambda_0 = 0.1$ m, $N_{\lambda_0} = 10$, $L = 50$ m, $\epsilon_{r1} = 70.4 + 40.6j$, $u_{10} = 5$ m/s, $\theta_{inc} = 30^o$ and the surface height spectrum is that of Elfouhaily et al. [ELF 97]*

As shown previously, as the wind speed and/or the frequency increases, the number of unknowns increases significantly, and then the impedance matrix can become difficult to invert from a standard algorithm, such as LU, with a standard computer. Then, specific algorithms are applied to overcome this issue. For more details, see [TSA 93a, TSA 93b, TSA 95] for the banded matrix iterative method/canonical grid (BMIA/CAG), [KAP 96, ADA 96, HOL 98, IOD 02] for the forward–backward (FB) and [CHO 02, CHO 00, TOR 00, TOR 02] for the FB spectral acceleration (FB-SA). These methods allow us to reduce the memory space requirement and the complexity of the impedance matrix inversion. In fact, this operation is not made explicitly, and the matrix-vector product $\bar{Z}^{-1}b$ is substituted for a succession of the matrix-vector products. It is also worth reading the topical reviews in [WAR 01, SAI 01].

2.5. FB method

In this section, the FB method is tested on a random rough surface, which was discussed in detail in Chapter 1.

Figure 2.28 shows the NRCS in dB scale computed from an LU (RCS_{LU}) inversion and the FB method (RCS_{FB}) versus the observation angle θ_{sca}. The simulation parameters are $\lambda_0 = 1$ m, $N_{\lambda_0} = 10$, $L = 300\lambda_0$, $\sigma_z = 0.5\lambda_0$, $L_c = 5\lambda_0$, $\theta_{\text{inc}} = 30^o$ and the surface height autocorrelation function is Gaussian: (a) Dirichlet boundary condition, (b) Neumann boundary condition, (c) Dielectric case, TE polarization and $\epsilon_{r1} = 2 + 0.1j$, and (d) Dielectric case, TM polarization and $\epsilon_{r1} = 2 + 0.1j$.

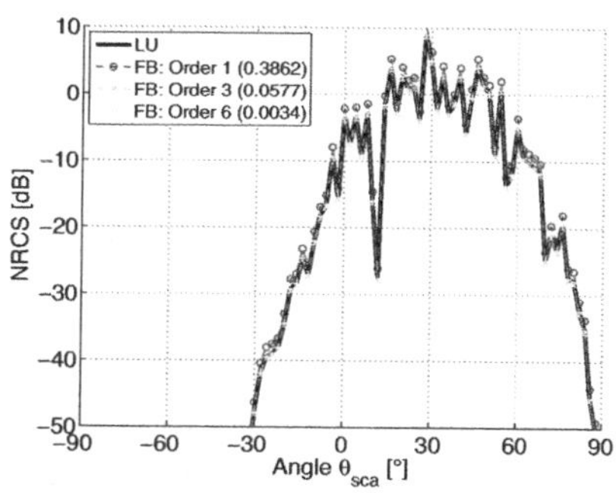

a) Dirichlet case (TE)

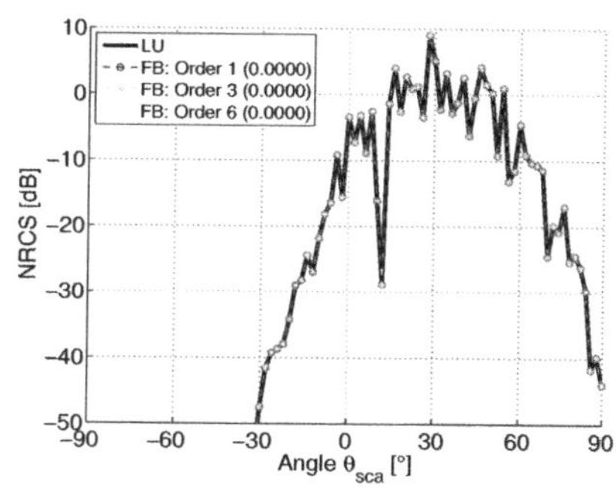

b) Neumann case (TM)

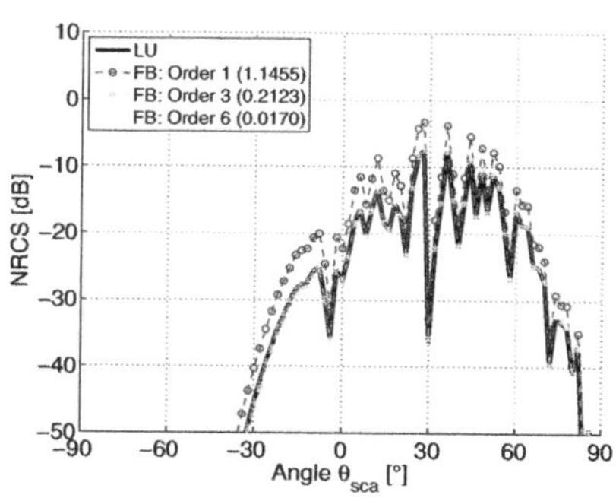

c) Dielectric case and TE

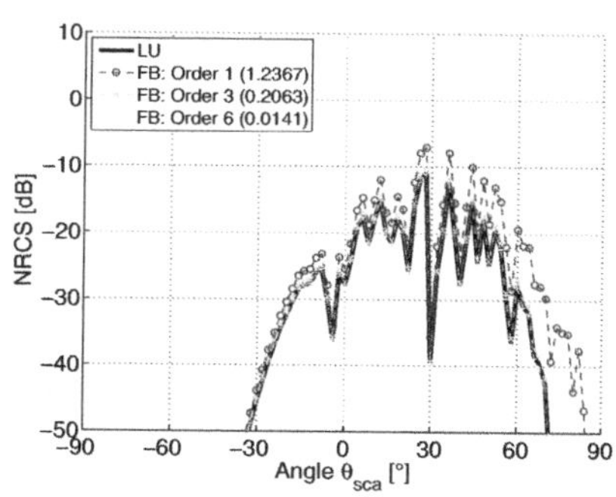

d) Dielectric case and TM

Figure 2.28. *NRCS in dB scale versus the observation angle θ_{sca} computed from an LU inversion and the FB method. $\lambda_0 = 1$ m, $N_{\lambda_0} = 10$, $L = 300\lambda_0$, $\sigma_z = 0.5\lambda_0$, $L_c = 5\lambda_0$, $\theta_{inc} = 30^o$ and the surface height autocorrelation function is Gaussian. a) Dirichlet boundary condition; b) Neumann boundary condition; c) dielectric case, TE polarization and $\epsilon_{r1} = 2 + 0.1j$, and d) dielectric case, TM polarization and $\epsilon_{r1} = 2 + 0.1j$*

In Figure 2.28, for a given FB order, the value in parentheses is the relative residual error defined as:

$$\epsilon_{\text{FB}} = \frac{\text{norm}_{\theta_{\text{sca}} \in [-\pi/2;\pi/2]} \left(\text{NRCS}_{\text{FB}} - \text{NRCS}_{\text{LU}}\right)}{\text{norm}_{\theta_{\text{sca}} \in [-\pi/2;\pi/2]} \left(\text{NRCS}_{\text{LU}}\right)}, \quad [2.80]$$

where the "norm" stands for the norm two.

The results show that the FB converges rapidly for the TE polarization and very rapidly for the TM polarization and the convergence is slower for a dielectric surface. It was shown in [BOU 08, KUB 08] that the convergence order of the FB method is quite insensitive to the height variance σ_z^2 and the incidence angle θ_{inc}.

For a dielectric rough surface, Iodice [IOD 02] studied in detail the convergence of the FB against the choice of the height autocorrelation function (HAF). For a Gaussian HAF, the FB always converges, whereas for an exponential HAF (with the same correlation length and variance height as the Gaussian case), the FB may fail for very rough surfaces. Déchamps *et al.* [DÉC 07b] mathematically showed that the FB method converges if the spectral radius (i.e. the modulus of the eigenvalue that has the highest modulus) of the characteristic matrix $\bar{\boldsymbol{M}}_{c,\text{FB}}$ is strictly smaller than one, where $\bar{\boldsymbol{M}}_{c,\text{FB}}$ is expressed as:

$$\bar{\boldsymbol{M}}_{c,\text{FB}} = \left(\bar{\boldsymbol{Z}}_{\text{Diag}} + \bar{\boldsymbol{Z}}_{\text{Forw}}\right)^{-1} \bar{\boldsymbol{Z}}_{\text{Forw}} \left(\bar{\boldsymbol{Z}}_{\text{Diag}} + \bar{\boldsymbol{Z}}_{\text{Back}}\right)^{-1} \bar{\boldsymbol{Z}}_{\text{Back}}. \quad [2.81]$$

Figure 2.29 shows the same variation as in Figure 2.28, but for a sea surface. The simulation parameters are $\lambda_0 = 0.1$ m, $N_{\lambda_0} = 10$, $L = 50$ m, $\epsilon_{r1} = 70.4 + 40.6j$, $u_{10} = 5$ m/s, $\theta_{\text{inc}} = 30^{\text{o}}$ and the surface height spectrum is that of Elfouhaily *et al.* [ELF 97]. The IBC approximation is applied to calculate the currents on the surface. The FB method converges more rapidly for the TM polarization.

The FB method accelerates the computation of the surface currents in comparison to a direct LU inversion, but the memory requirement is the same for the both methods, since the elements of the impedance matrix must be stored. From the FB algorithmic scheme, it is possible to compute for each iteration and for a given component of the current vector, only a row of $\bar{\boldsymbol{Z}}$. Then, the memory requirement becomes N instead of N^2, but the computing time increases, because this row must be recalculated for each iteration. To solve this issue, the FB combined with the spectral-acceleration (FB-SA) [CHO 02, CHO 00, TOR 00, TOR 02] has been developed to both reduce the memory requirement and the computing time (complexity $\mathcal{O}(N)$ instead of $\mathcal{O}(N^2)$ with the FB). Its principle is to split up the impedance matrix into strong and weak interactions. Then, the strong interactions are calculated from the FB, whereas the weak interactions are computed from the SA. Since the limit distance between the strong and weak interactions $x_{\rm SA}$ is much shorter than the surface length, the complexity of the method is that of SA, that is $\mathcal{O}(N)$. The BMIA-CAG [TSA 93a, TSA 93b, TSA 95] use the same process. The FB-SA is very efficient for a sea surface, since $x_{\rm SA} = 0.02u_{10}^2$ [KUB 10c, KUB 10b], and as well as for a Gaussian autocorrelation function, since $x_{\rm SA} = 3L_c$ [BOU 08, KUB 08]. The FB-SA has also shown its efficiency for the diffraction from the Earth in High-Frequency (HF) band ($[3-30]$ MHz) [BOU 11a, BOU 11b].

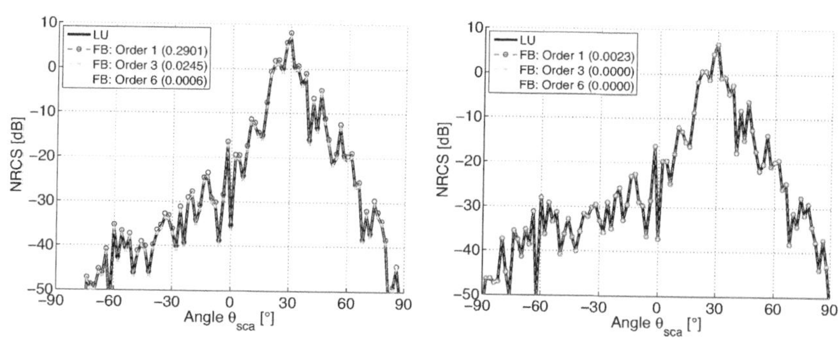

a) TE polarization   b) TM polarization

Figure 2.29. *Same variation as in Figure 2.28, but for a sea surface. $\lambda_0 = 0.1$ m, $N_{\lambda_0} = 10$, $L = 50$ m, $\epsilon_{r1} = 70.4 + 40.6j$, $u_{10} = 5$ m/s, $\theta_{inc} = 30^o$ and the surface height spectrum is that of Elfouhaily et al. [ELF 97]*

2.6. Conclusion

In this chapter, from the MoM, the currents on the surface and then the corresponding RCSs are computed. The results are then compared with the exact solution of an infinite circular cylinder, leading to a very good agreement between the two methods. In addition, for a perfectly conducting elliptical cylinder, a perfectly conducting plate and a rough dielectric surface, the PO approximation has been compared with the MoM.

Here, the scatterer is assumed to be in free space. In many scattering problems, the scatterer is often in the presence of another scatterer. For instance, this topic concerns the scattering of an object near (below or above) a random rough surface or the scattering from a stack of two one-dimensional rough interfaces separating homogeneous media. Chapters 3 and 4 solve this issue with the MoM.

3

Scattering from Two Illuminated Scatterers

3.1. Introduction

The study of scattering from several objects or several objects near a surface (smooth or rough) is an active research area. Many applications are concerned by these works, for instance in the remote sensing domain, for the detection of ocean ice, sand cover of arid regions, or oil slicks on the ocean. Also in the optics domain, for instance, for optical studies of thin films and coated surfaces and in the treatment of antireflection coatings. When the objects are circular cylinders near a smooth surface [SOM 64, YOU 88, VAL 94, MAD 95, VID 97, HEN 07, AHM 08, LEE 11, PAW 11, FIA 12, PAW 12], exact closed-form expressions of the scattered field can be obtained, but the complexity of programming is high. When the shape of the cylinder is elliptic [TSO 10, ZOU 11, ZOU 13], it is also possible to have a closed form expression of the scattered field by introducing the Mathieu functions [ABR 70], but these functions are very difficult to program, unlike the well-known Bessel functions, used for circular cylinders. References cited above can be referred for an exhaustive review of these issues. A means of validating the resulting formula is to compare it with another exact method, such as the method of moments (MoM), presented in this chapter. In addition, the advantage of the MoM is that it can be used for any incident wave, instead of a plane wave, which is often used in the above references. The MatLab codes provided with this book can be used by researchers or engineers to validate their models. When the surface is rough, some asymptotic models and exact numerical methods have been investigated

[PIN 99, JOH 02, WAN 03, LIU 04, YE 06, DON 07, YE 07] (see also references therein for an exhaustive review).

For the problem of scattering from a *single* electrically large or multi-scale (like the sea) rough surface (i.e. many unknowns), starting from the MoM, exact, fast numerical methods have been investigated to solve this issue (see Chapter 2). For instance, we can quote the banded-matrix-iterative-approach/canonical grid (BMIA-CAG) of Tsang *et al.* [TSA 93a, TSA 93b, TSA 95] of complexity $\mathcal{O}(N \log N)$, the forward–backward (FB) method of Holliday *et al.* [KAP 96, ADA 96, HOL 98, IOD 02] of complexity $\mathcal{O}(N^2)$ and the accelerated FB spectral acceleration (FB-SA) version of Chou *et al.* [CHO 02, CHO 00, TOR 00, TOR 02, PIN 99] of complexity $\mathcal{O}(N)$, in which N is the number of unknowns on the surface.

In the same spirit, from the MoM, Kubické *et al.* have developed the extended propagation-inside-layer expansion (E-PILE) method to explain the case of two illuminated scatterers and applied it to an object located above a rough surface [KUB 08, KUB 10a, KUB 10c, KUB 10b]. In [KUB 08], [KUB 10c] and [KUB 10b], E-PILE is combined with the FB-SA to calculate the local interactions on the rough surface. Moreover, in [KUB 10a], the E-PILE is combined with both the FB-SA and the physical optics (PO) approximation up to second order to calculate the local interactions on the object (a cross). As a result, since the number of unknowns on the surface N_1 is much greater than that of the object N_2, the complexity of the method is reduced to $\mathcal{O}(N_1)$ instead of $\mathcal{O}([N_1 + N_2]^3)$, if a direct lower–upper (LU) inversion is applied to inverse the impedance matrix of the two scatterers.

The aim of this chapter is to present the efficient E-PILE method and to explain how the calculation of the local interactions on each scatterer can be accelerated by the PO and FB methods, which are valid for a single scatterer. Chapter 4 will present the PILE method, for which only one scatterer is illuminated from the incident wave, and numerical results will be shown for a coated cylinder, an object below a rough surface and a rough layer.

The following section presents the integral equations and the MoM, in which the MoM is generalized to $P > 2$ scatterers. In section 3.3, the E-PILE algorithm is presented and section 3.4 presents how PO and FB can be combined with E-PILE. Several numerical examples are also analyzed, like the scattering from two cylinders, from two PC plates, from P plates and cylinders, and an object above a rough surface.

3.2. Integral equations and method of moments

3.2.1. *Integral equations for two scatterers*

The problem to be solved is illustrated in Figure 3.1. A source is defined in the medium Ω_0 of permittivity ϵ_0. This medium contains two scatterers: scatterer 1 of medium Ω_1 and of permittivity ϵ_1, and scatterer 2 of medium Ω_2 and of permittivity ϵ_2. The surface S_i $(i = \{1, 2\})$ bounds Ω_i; its normal, $\hat{\boldsymbol{n}}_i$, points toward Ω_0. S_∞ is the surface bounding the space Ω_0; its normal, $\hat{\boldsymbol{n}}_\infty$, points toward outside Ω_0.

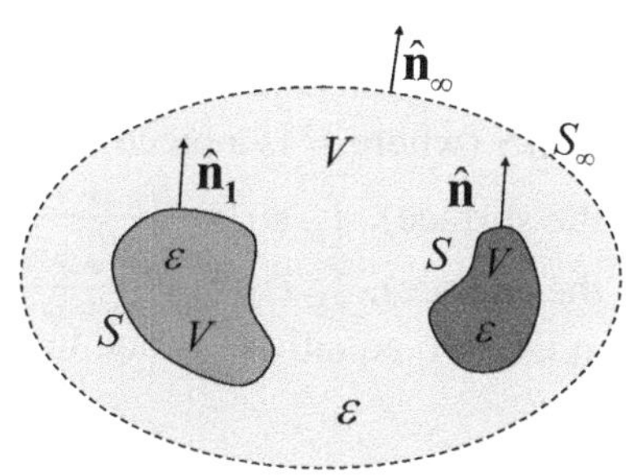

Figure 3.1. *Scattering from two scatterers: integral equations*

From the Helmoltz wave propagation equation [1.13] in each medium, from the definition of the Green function [1.14] and from the far radiation condition, the use of the Green theorem for two scatterers leads to [WAN 03]:

$$\begin{aligned}\begin{cases}\psi_0(\boldsymbol{r}') \text{ if } \boldsymbol{r}' \in \Omega_0 \\ 0 \qquad \text{otherwise}\end{cases} &= \psi_{\text{inc}}(\boldsymbol{r}') \\ &+ \int_{S_1} \left[\psi_1(\boldsymbol{r}) \frac{\partial g_0(\boldsymbol{r}, \boldsymbol{r}')}{\partial n} - g_0(\boldsymbol{r}, \boldsymbol{r}') \frac{\partial \psi_1(\boldsymbol{r})}{\partial n}\right] dS \\ &+ \int_{S_2} \left[\psi_2(\boldsymbol{r}) \frac{\partial g_0(\boldsymbol{r}, \boldsymbol{r}')}{\partial n} - g_0(\boldsymbol{r}, \boldsymbol{r}') \frac{\partial \psi_2(\boldsymbol{r})}{\partial n}\right] dS, \quad [3.1]\end{aligned}$$

where ψ_{inc} is the incident field defined inside the medium Ω_0. In equation [3.1], if $\boldsymbol{r}' \to S_1^+$ (case $\boldsymbol{r}' \in \Omega_0$), then:

$$\psi_{\text{inc}}(\boldsymbol{r}') = \underbrace{+\frac{1}{2}\psi_1(\boldsymbol{r}') - \fint_{S_1} \frac{\partial g_0(\boldsymbol{r},\boldsymbol{r}')}{\partial n}\psi_1(\boldsymbol{r})dS}_{\Rightarrow \bar{\boldsymbol{A}}_1} + \underbrace{\int_{S_1} g_0(\boldsymbol{r},\boldsymbol{r}')\frac{\partial \psi_1(\boldsymbol{r})}{\partial n}dS}_{\Rightarrow \bar{\boldsymbol{B}}_1}$$
$$\underbrace{-\int_{S_2} \frac{\partial g_0(\boldsymbol{r},\boldsymbol{r}')}{\partial n}\psi_2(\boldsymbol{r})dS}_{\Rightarrow \bar{\boldsymbol{A}}_{21}} + \underbrace{\int_{S_2} g_0(\boldsymbol{r},\boldsymbol{r}')\frac{\partial \psi_2(\boldsymbol{r})}{\partial n}dS}_{\Rightarrow \bar{\boldsymbol{B}}_{21}}. \tag{3.2}$$

The symbol $\fint$ stands for the Cauchy principal value, which means that the case $\boldsymbol{r} = \boldsymbol{r}'$ is not accounted for in the calculation of the integral. In addition, letting $\int_{S_1} dS = \fint dS + \int_P dS$ (where P is a piece), it is important to note that for $\boldsymbol{r} = \boldsymbol{r}' \in S_1^+$ (above the surface), $\int_P \psi_1(\boldsymbol{r})\frac{\partial g_0(\boldsymbol{r},\boldsymbol{r}')}{\partial n}dS = +\psi_1(\boldsymbol{r}')/2$, and for $\boldsymbol{r} = \boldsymbol{r}' \in S_1^-$ (below the surface), $\int_P \psi_1(\boldsymbol{r})\frac{\partial g_0(\boldsymbol{r},\boldsymbol{r}')}{\partial n}dS = -\psi_1(\boldsymbol{r}')/2$. The difference in sign between the two equations is due to the sense of the normal to the surface S_1.

In equation [3.1], if $\boldsymbol{r}' \to S_2^+$ (case $\boldsymbol{r}' \in \Omega_0$), the use of the same method leads to:

$$\psi_{\text{inc}}(\boldsymbol{r}') = \underbrace{+\frac{1}{2}\psi_2(\boldsymbol{r}') - \fint_{S_2} \frac{\partial g_0(\boldsymbol{r},\boldsymbol{r}')}{\partial n}\psi_2(\boldsymbol{r})dS}_{\Rightarrow \bar{\boldsymbol{A}}_2} + \underbrace{\int_{S_2} g_0(\boldsymbol{r},\boldsymbol{r}')\frac{\partial \psi_2(\boldsymbol{r})}{\partial n}dS}_{\Rightarrow \bar{\boldsymbol{B}}_2}$$
$$\underbrace{-\int_{S_1} \frac{\partial g_0(\boldsymbol{r},\boldsymbol{r}')}{\partial n}\psi_1(\boldsymbol{r})dS}_{\Rightarrow \bar{\boldsymbol{A}}_{12}} + \underbrace{\int_{S_1} g_0(\boldsymbol{r},\boldsymbol{r}')\frac{\partial \psi_1(\boldsymbol{r})}{\partial n}dS}_{\Rightarrow \bar{\boldsymbol{B}}_{12}}. \tag{3.3}$$

The four unknowns to be determined are the currents $\{\psi_i, \partial\psi_i/\partial n\}$ on the surfaces $\{S_i\}$ ($i = \{1,2\}$). Two additional equations are required. Applying the Green theorem in medium Ω_i, we have [TSA 00] (or equation [1.20]):

$$\begin{cases} 0 & = \int_{S_i} \left[\psi_i(\boldsymbol{r}) \frac{\partial g_i(\boldsymbol{r},\boldsymbol{r}')}{\partial n} - g_i(\boldsymbol{r},\boldsymbol{r}') \frac{\partial \psi_i(\boldsymbol{r})}{\partial n} \right] dS \text{ if } \boldsymbol{r}' \notin \Omega_i \\ -\psi_i(\boldsymbol{r}') & = \int_{S_i} \left[\psi_i(\boldsymbol{r}) \frac{\partial g_i(\boldsymbol{r},\boldsymbol{r}')}{\partial n} - g_i(\boldsymbol{r},\boldsymbol{r}') \frac{\partial \psi_i(\boldsymbol{r})}{\partial n} \right] dS \text{ if } \boldsymbol{r}' \in \Omega_i \end{cases} . \quad [3.4]$$

From the boundary conditions on scatterer i expressed as:

$$\begin{cases} \psi_0(\boldsymbol{r}) = \psi_i(\boldsymbol{r}) \\ \dfrac{\partial \psi_0(\boldsymbol{r})}{\partial n} = \rho_{0i} \dfrac{\partial \psi_i(\boldsymbol{r})}{\partial n}, \end{cases} \quad [3.5]$$

where $\rho_{0i} = 1$ for the transverse electric (TE) polarization and $\rho_{0i} = \epsilon_0/\epsilon_i$ for the transverse magnetic (TM) polarization, we have for $\boldsymbol{r}' \to S_1$ ($\boldsymbol{r}' \in \Omega_1$):

$$0 = \underbrace{-\frac{1}{2}\psi_1(\boldsymbol{r}') - \fint_{S_1} \psi_1(\boldsymbol{r}) \frac{\partial g_1(\boldsymbol{r},\boldsymbol{r}')}{\partial n} dS}_{\bar{\boldsymbol{C}}_1} + \frac{1}{\rho_{01}} \underbrace{\int_{S_1} g_1(\boldsymbol{r},\boldsymbol{r}') \frac{\partial \psi_1(\boldsymbol{r})}{\partial n} dS}_{\bar{\boldsymbol{D}}_1}, \quad [3.6]$$

and for $\boldsymbol{r}' \to S_2$ ($\boldsymbol{r}' \in \Omega_2$):

$$0 = \underbrace{-\frac{1}{2}\psi_2(\boldsymbol{r}') - \fint_{S_2} \psi_2(\boldsymbol{r}) \frac{\partial g_2(\boldsymbol{r},\boldsymbol{r}')}{\partial n} dS}_{\bar{\boldsymbol{C}}_2} + \frac{1}{\rho_{02}} \underbrace{\int_{S_2} g_2(\boldsymbol{r},\boldsymbol{r}') \frac{\partial \psi_2(\boldsymbol{r})}{\partial n} dS}_{\bar{\boldsymbol{D}}_2}. \quad [3.7]$$

From equations [3.2], [3.3], [3.6] and [3.7], the unknowns on the surfaces S_1 and S_2 can be computed from the MoM.

3.2.2. *Method of moments for two scatterers*

In this section, the MoM is applied with the point-matching method and the pulse basis function to convert the integral equations [3.2], [3.3], [3.6] and [3.7] into a linear system. This method is thoroughly explained in section 1.3.

Equations [3.2], [3.3], [3.6] and [3.7] lead to a coupled system. From the MoM, the equations are discretized on the surface of each scatterer, leading to the linear system $\bar{\boldsymbol{Z}}\boldsymbol{X} = \boldsymbol{b}$. The unknown vector $\boldsymbol{X}$ is then:

$$\boldsymbol{X} = \begin{bmatrix} \boldsymbol{X}_1 \\ \boldsymbol{X}_2 \end{bmatrix}, \quad [3.8]$$

where the components of the vectors $\boldsymbol{X}_1$ and $\boldsymbol{X}_2$ are the currents discretized on the surface S_1 and S_2, respectively. They are written as:

$$\boldsymbol{X}_1 = \left[\psi_1(\boldsymbol{r}_1) \ldots \psi_1(\boldsymbol{r}_{N_1}) \ \frac{\partial \psi_1(\boldsymbol{r}_1)}{\partial n} \ldots \frac{\partial \psi_1(\boldsymbol{r}_{N_1})}{\partial n} \right]^{\mathrm{T}} \quad \boldsymbol{r}_{p \in [1;N_1]} \in S_1, \quad [3.9]$$

$$\boldsymbol{X}_2 = \left[\psi_2(\boldsymbol{r}_1) \ldots \psi_2(\boldsymbol{r}_{N_2}) \ \frac{\partial \psi_2(\boldsymbol{r}_2)}{\partial n} \ldots \frac{\partial \psi_2(\boldsymbol{r}_{N_2})}{\partial n} \right]^{\mathrm{T}} \quad \boldsymbol{r}_{p \in [1;N_2]} \in S_2, [3.10]$$

where the symbol T stands for transpose, and N_i is the number of samples on the surface S_i. The length of the vector $\boldsymbol{X}_i$ is then $2N_i$.

The vector $\boldsymbol{b}$ of length $2(N_1 + N_2)$ is the incident field discretized on the surfaces S_1 and S_2. It is defined as:

$$\boldsymbol{b} = \begin{bmatrix} \boldsymbol{b}_1 \\ \boldsymbol{b}_2 \end{bmatrix} = \Bigg[\underbrace{\psi_{\text{inc}}(\boldsymbol{r}_1) \ldots \psi_{\text{inc}}(\boldsymbol{r}_{N_1}) \ \underbrace{0 \ldots 0}_{N_1 \text{ times}}}_{\boldsymbol{b}_1^{\mathrm{T}}, \boldsymbol{r} \in S_1} \ \underbrace{\psi_{\text{inc}}(\boldsymbol{r}_1) \ldots \psi_{\text{inc}}(\boldsymbol{r}_{N_2}) \ \underbrace{0 \ldots 0}_{N_2 \text{ times}}}_{\boldsymbol{b}_2^{\mathrm{T}}, \boldsymbol{r} \in S_2} \Bigg]^{\mathrm{T}}. \quad [3.11]$$

The impedance matrix of size $2(N_1 + N_2) \times 2(N_1 + N_2)$ is then:

$$\bar{\boldsymbol{Z}} = \begin{bmatrix} \bar{\boldsymbol{A}}_1 & \bar{\boldsymbol{B}}_1 & \bar{\boldsymbol{A}}_{21} & \bar{\boldsymbol{B}}_{21} \\ \bar{\boldsymbol{C}}_1 & \dfrac{1}{\rho_{01}}\bar{\boldsymbol{D}}_1 & \bar{0} & \bar{0} \\ \bar{\boldsymbol{A}}_{12} & \bar{\boldsymbol{B}}_{12} & \bar{\boldsymbol{A}}_2 & \bar{\boldsymbol{B}}_2 \\ \bar{0} & \bar{0} & \bar{\boldsymbol{C}}_2 & \dfrac{1}{\rho_{02}}\bar{\boldsymbol{D}}_2 \end{bmatrix} = \begin{bmatrix} \bar{\boldsymbol{Z}}_1 & \bar{\boldsymbol{Z}}_{21} \\ \bar{\boldsymbol{Z}}_{12} & \bar{\boldsymbol{Z}}_2 \end{bmatrix}, \quad [3.12]$$

where

$$\bar{\boldsymbol{Z}}_1 = \begin{bmatrix} \bar{\boldsymbol{A}}_1 & \bar{\boldsymbol{B}}_1 \\ \bar{\boldsymbol{C}}_1 & \dfrac{1}{\rho_{01}}\bar{\boldsymbol{D}}_1 \end{bmatrix}, \quad \bar{\boldsymbol{Z}}_2 = \begin{bmatrix} \bar{\boldsymbol{A}}_2 & \bar{\boldsymbol{B}}_2 \\ \bar{\boldsymbol{C}}_2 & \dfrac{1}{\rho_{02}}\bar{\boldsymbol{D}}_2 \end{bmatrix}, \quad [3.13]$$

and

$$\bar{\boldsymbol{Z}}_{21} = \begin{bmatrix} \bar{\boldsymbol{A}}_{21} & \bar{\boldsymbol{B}}_{21} \\ \bar{0} & \bar{0} \end{bmatrix}, \quad \bar{\boldsymbol{Z}}_{12} = \begin{bmatrix} \bar{\boldsymbol{A}}_{12} & \bar{\boldsymbol{B}}_{12} \\ \bar{0} & \bar{0} \end{bmatrix}. \qquad [3.14]$$

The impedance matrix $\bar{\boldsymbol{Z}}_i$ of size $2N_i \times 2N_i$ is the impedance matrix of the scatterer i in free space.

The matrix $\bar{\boldsymbol{A}}_i$ of size $N_i \times N_i$ is the matrix of a Neumann boundary condition problem expressed from equation [1.55] with $\Delta_n \to v_n|\Delta_n|$ for any orientation of the normal to the surface. Its elements are:

$$A_{i,mn} = \begin{cases} -\dfrac{jk_0 v_n|\Delta_n|}{4} \dfrac{H_1^{(1)}(k_0\|\boldsymbol{r}_n - \boldsymbol{r}_m\|)}{\|\boldsymbol{r}_n - \boldsymbol{r}_m\|} \\ \qquad \times [\gamma_n(x_n - x_m) - (z_n - z_m)] & \text{for } m \neq n, \\ +\dfrac{1}{2} - \dfrac{v_n|\Delta_n|}{4\pi} \dfrac{\gamma'(x_m)}{1+\gamma^2(x_m)} & \text{for } m = n \end{cases} \qquad [3.15]$$

where $(x_n, z_n) \in S_i$, $(x_m, z_m) \in S_i$, $\gamma = dz/dx$, $\|\boldsymbol{r}_n - \boldsymbol{r}_m\| = \sqrt{(x_n - x_m)^2 + (z_n - z_m)^2}$, $v = \hat{\boldsymbol{n}} \cdot \hat{\boldsymbol{z}}$ is the sense of the normal to the surface, Δ_n is the sampling step, $\mathrm{H}_0^{(1)}$ is the zeroth-order Hankel function of the first kind and $\mathrm{H}_1^{(1)}$ is its derivative.

The matrix $\bar{\boldsymbol{B}}_i$ of size $N_i \times N_i$ is the matrix of a Dirichlet boundary condition problem expressed from equation [1.52] with $\Delta_n \to |\Delta_n|$. Its elements are:

$$B_{i,mn} = \frac{j|\Delta_n|\sqrt{1+\gamma_n^2}}{4} \begin{cases} H_0^{(1)}(k_0\,\|\boldsymbol{r}_n - \boldsymbol{r}_m\|) & \text{for } m \neq n \\ 1 + \dfrac{2j}{\pi} \ln\left(0.164 k_0 \sqrt{1+\gamma_n^2}|\Delta_n|\right) & \text{for } m = n \end{cases}, \qquad [3.16]$$

where $(x_n, z_n) \in S_i$ and $(x_m, z_m) \in S_i$.

The matrix $\bar{\boldsymbol{C}}_i$ is similar to the matrix $\bar{\boldsymbol{A}}_i$ but the Cauchy principal value is $-1/2$ instead of $+1/2$ for the calculation of the diagonal elements $C_{i,nn}$. In addition, the matrices $\bar{\boldsymbol{C}}_i$ and $\bar{\boldsymbol{D}}_i$ are obtained from the matrices $\bar{\boldsymbol{A}}_i$ and $\bar{\boldsymbol{B}}_i$ by substituting k_0 for k_i.

The matrices $\bar{\boldsymbol{Z}}_{21}$ (of size $2N_1 \times 2N_2$) and $\bar{\boldsymbol{Z}}_{12}$ (of size $2N_2 \times 2N_1$) can be interpreted as coupling matrices. For instance, from equation [3.3], the matrices $\bar{\boldsymbol{A}}_{12}$ and $\bar{\boldsymbol{B}}_{12}$ multiplied by the vector containing the currents on S_1 (ψ_1 and $\partial\psi_1/\partial n$, respectively) give a scattered field calculated on scatterer 2. Then, the matrices $\bar{\boldsymbol{A}}_{12}$ and $\bar{\boldsymbol{B}}_{12}$ propagate the currents from scatterer 1 to scatterer 2. It is then a coupling matrix. The same remark holds for the matrices $\bar{\boldsymbol{A}}_{21}$ and $\bar{\boldsymbol{B}}_{21}$. These matrices are expressed from the matrices $\bar{\boldsymbol{A}}_i$ and $\bar{\boldsymbol{B}}_i$, with the main difference that the case $n = m$ does not occur since a point of scatterer 1 always points onto a point of scatterer 2 and vice versa. Their elements are expressed as:

$$\left\{\begin{aligned} A_{12,mn} &= -\frac{jk_0v_{1,n}|\Delta_{1,n}|}{4}\frac{H_1^{(1)}\left(k_0\|\boldsymbol{r}_{1,n}-\boldsymbol{r}_{2,m}\|\right)}{\|\boldsymbol{r}_{1,n}-\boldsymbol{r}_{2,m}\|} \\ &\quad\times\left[\gamma_{1,n}(x_{1,n}-x_{2,m})-(z_{1,n}-z_{2,m})\right] \\ A_{21,mn} &= -\frac{jk_0v_{2,n}|\Delta_{2,n}|}{4}\frac{H_1^{(1)}\left(k_0\|\boldsymbol{r}_{2,n}-\boldsymbol{r}_{1,m}\|\right)}{\|\boldsymbol{r}_{2,n}-\boldsymbol{r}_{1,m}\|} \\ &\quad\times\left[\gamma_{2,n}(x_{2,n}-x_{1,m})-(z_{2,n}-z_{1,m})\right] \\ B_{12,mn} &= \frac{j|\Delta_{1,n}|\sqrt{1+\gamma_{1,n}^2}}{4}H_0^{(1)}\left(k_0\|\boldsymbol{r}_{1,n}-\boldsymbol{r}_{2,m}\|\right) \\ B_{21,mn} &= \frac{j|\Delta_{2,n}|\sqrt{1+\gamma_{2,n}^2}}{4}H_0^{(1)}\left(k_0\|\boldsymbol{r}_{2,n}-\boldsymbol{r}_{1,m}\|\right) \end{aligned}\right., \qquad [3.17]$$

for any (n, m). The subscripts "$1, n$" and "$2, m$" refer to a point on scatterer 1 and 2, respectively.

From the knowledge of the currents $\{\psi_i, \partial\psi_i/\partial n\}$ on the scatterers, the scattered field $\psi_{\text{sca},\nu}$ in medium $\boldsymbol{r}' \in \Omega_\nu$ ($\nu = \{0, 1, 2\}$) is computed from the Huygens' principle as:

$$\left\{\begin{aligned} \psi_{\text{sca},0}(\boldsymbol{r}') &= \sum_{p=1}^{p=2}\int_{S_p}\left[\psi_p(\boldsymbol{r})\frac{\partial g_0(\boldsymbol{r},\boldsymbol{r}')}{\partial n}-g_0(\boldsymbol{r},\boldsymbol{r}')\frac{\partial\psi_p(\boldsymbol{r})}{\partial n}\right]dS \\ \psi_{\text{sca},1}(\boldsymbol{r}') &= -\int_{S_1}\left[\psi_1(\boldsymbol{r})\frac{\partial g_1(\boldsymbol{r},\boldsymbol{r}')}{\partial n}-g_1(\boldsymbol{r},\boldsymbol{r}')\frac{\partial\psi_1(\boldsymbol{r})}{\partial n}\right]dS \\ \psi_{\text{sca},2}(\boldsymbol{r}') &= -\int_{S_2}\left[\psi_2(\boldsymbol{r})\frac{\partial g_2(\boldsymbol{r},\boldsymbol{r}')}{\partial n}-g_2(\boldsymbol{r},\boldsymbol{r}')\frac{\partial\psi_2(\boldsymbol{r})}{\partial n}\right]dS \end{aligned}\right. . \qquad [3.18]$$

3.2.2.1. *Case of perfectly conducting scatterers*

For a perfectly conducting (PC) scatterer i, the impedance matrix Z_i can be simplified. For the TE polarization (Dirichlet boundary condition), ψ_i on the surface vanishes and the only unknown on the surface is $\partial\psi_i/\partial n$. Then:

$$\text{TE: } \bar{\boldsymbol{Z}}_i = \bar{\boldsymbol{B}}_i, \; \boldsymbol{X}_i = \frac{\partial\psi_i}{\partial n}. \tag{3.19}$$

For the TM polarization (Neumann boundary condition), $\partial\psi_i/\partial n$ on the surface vanishes and the only unknown on the surface is ψ_i. Then:

$$\text{TM: } \bar{\boldsymbol{Z}}_i = \bar{\boldsymbol{A}}_i, \; \boldsymbol{X}_i = \psi_i. \tag{3.20}$$

In addition, if scatterer 1 is PC and scatterer 2 is dielectric, then:

$$\begin{cases} \text{TE: } \bar{\boldsymbol{Z}}_1 = \bar{\boldsymbol{B}}_1 \; \bar{\boldsymbol{Z}}_{21} = \left[\bar{\boldsymbol{A}}_{21} \; \bar{\boldsymbol{B}}_{21}\right] \; \bar{\boldsymbol{Z}}_{12} = \begin{bmatrix} \bar{\boldsymbol{B}}_{12} \\ \bar{\boldsymbol{0}} \end{bmatrix} \\ \\ \text{TM: } \bar{\boldsymbol{Z}}_1 = \bar{\boldsymbol{A}}_1 \; \bar{\boldsymbol{Z}}_{21} = \left[\bar{\boldsymbol{A}}_{21} \; \bar{\boldsymbol{B}}_{21}\right] \; \bar{\boldsymbol{Z}}_{12} = \begin{bmatrix} \bar{\boldsymbol{A}}_{12} \\ \bar{\boldsymbol{0}} \end{bmatrix} \end{cases}, \tag{3.21}$$

and the matrix $\bar{\boldsymbol{Z}}_2$ is expressed from equation [3.13]. In addition, if scatterer 2 is PC, then $\bar{\boldsymbol{Z}}_i = \bar{\boldsymbol{B}}_i$ ($i = \{1, 2\}$) and $\bar{\boldsymbol{Z}}_{ij} = \bar{\boldsymbol{B}}_{ij}$ ($j = \{1, 2\}$ with $j \neq i$) for the TE polarization, whereas $\bar{\boldsymbol{Z}}_i = \bar{\boldsymbol{A}}_i$ and $\bar{\boldsymbol{Z}}_{ij} = \bar{\boldsymbol{A}}_{ij}$ for the TM polarization.

If the scatterer 2 is PC and the scatterer 1 is dielectric, then the matrices are obtained from equation [3.21] by substituting subscripts $\{1, 12, 21\}$ for $\{2, 21, 12\}$, and $\bar{\boldsymbol{Z}}_1$ is expressed from equation [3.13].

3.2.2.2. *IBC approximation*

From section 1.4.4, if the scatterer i satisfies the Leontovitvh boundary condition or the impedance boundary condition (IBC), then:

$$\bar{\boldsymbol{Z}}_i = \alpha_i \bar{\boldsymbol{A}}_i + \beta_i \bar{\boldsymbol{B}}_i, \tag{3.22}$$

with

$$\begin{cases} \text{TE} : \beta_i = 1 & \alpha_i = \dfrac{j}{k_0}\sqrt{\dfrac{\epsilon_{r0}}{\epsilon_{ri}}} \quad \boldsymbol{X}_i = \dfrac{\partial \psi_i(\boldsymbol{r})}{\partial n} \quad \psi_i(\boldsymbol{r}) = \alpha_i \dfrac{\partial \psi_i(\boldsymbol{r})}{\partial n} \\ \text{TM} : \beta_i = \dfrac{k_0}{j}\sqrt{\dfrac{\epsilon_{r0}}{\epsilon_{ri}}} \quad \alpha_i = 1 & \boldsymbol{X}_i = \psi_i(\boldsymbol{r}) \quad \dfrac{\partial \psi_i(\boldsymbol{r})}{\partial n} = \beta_i \psi_0(\boldsymbol{r}) \end{cases}. \quad [3.23]$$

In addition, if the scatterer 1 satisfies the IBC approximation and the scatterer 2 is dielectric, from equation [3.21], the coupling matrix $\bar{\boldsymbol{Z}}_{21}$ remains unchanged and the coupling matrix $\bar{\boldsymbol{Z}}_{12}$ is:

$$\bar{\boldsymbol{Z}}_{12} = \begin{bmatrix} \alpha_1 \bar{\boldsymbol{A}}_{12} + \beta_1 \bar{\boldsymbol{B}}_{12} \\ \bar{\boldsymbol{0}} \end{bmatrix}. \quad [3.24]$$

If scatterers 1 and 2 satisfy the IBC approximation, then $\bar{\boldsymbol{Z}}_i = \alpha_i \bar{\boldsymbol{A}}_i + \beta_i \bar{\boldsymbol{B}}_i$ and $\bar{\boldsymbol{Z}}_{ij} = \alpha_i \bar{\boldsymbol{B}}_{ij} + \beta_i \bar{\boldsymbol{A}}_{ij}$.

3.2.2.3. *Numerical results: case of two elliptical cylinders*

Like for the scattering from a single circular cylinder (see section 2.3), the exact solution of the scattering from two cylinders can be obtained. Refer, for instance, reference [YOU 88]. Then, to apply the boundary condition on each cylinder, the coordinates of cylinder 1 are expressed from the coordinates of cylinder 2 via the Graf addition theorem [ABR 70] on the Bessel functions. Since the multiple reflections are accounted for, the complexity of programming increases significantly. In addition, the incident field is often considered as a plane wave. With the MoM, this is not a constraint and it can also be applied to an elliptical cylinder without modification of the MatLab code.

Figure 3.2 shows the modulus and the phase of the currents $\{\psi_i, \partial\psi_i/\partial n\}$ versus θ_i of two dielectric elliptical cylinders of semi-major $a_1 = a_2 = 2\lambda_0$ and semi-minor $b_1 = b_2 = \lambda_0$ axes, of centers $C_i = (\mp 2, 0)\lambda_0$ and of rotation angles $\alpha_i = \mp\pi/4$ ($\lambda_0 = 1$ m, $\epsilon_{r1} = \epsilon_{r2} = 4 + 0.05j$, $\psi_{\text{inc},0} = 1$, $\theta_{\text{inc}} = 0$ and TE polarization).

In addition, the currents are plotted when the coupling is neglected, which means that the coupling matrices $\bar{\boldsymbol{Z}}_{12}$ and $\bar{\boldsymbol{Z}}_{21}$ are taken as zeros. Figure 3.3a) shows the total field versus the abscissa x and the height z, and Figure 3.3b) shows the radar cross-section (RCS) in dBm scale versus the observation angle

θ_{sca}. For the cylinder i, $\theta_i = 0$ corresponds to the points of coordinates $(\mp 2 + \sqrt{2}, \mp\sqrt{2})\lambda_0$.

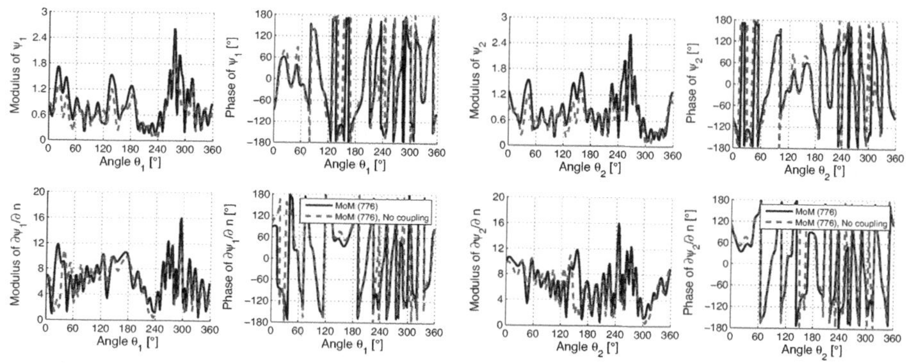

a) ψ_1 and $\partial\psi_1/\partial n$ b) ψ_2 and $\partial\psi_2/\partial n$

Figure 3.2. *Modulus and phase of the currents $\{\psi_i, \partial\psi_i/\partial n\}$ versus θ_i of two dielectric elliptical cylinders of semi-major $a_1 = a_2 = 2\lambda_0$ and semi-minor $b_1 = b_2 = \lambda_0$ axes, of centers $C_i = (\mp 2, 0)\lambda_0$ and of rotation angles $\alpha_i = \mp\pi/4$. $\lambda_0 = 1$ m, $\epsilon_{r1} = \epsilon_{r2} = 4 + 0.05j$, $\psi_{inc,0} = 1$, $\theta_{inc} = 0$ and TE polarization*

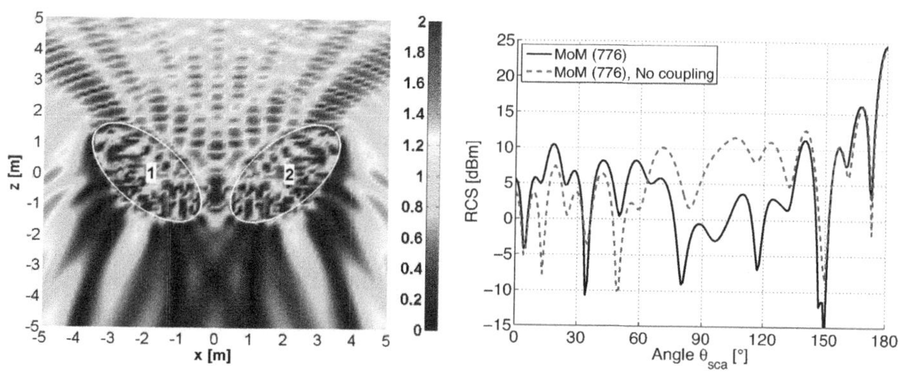

a) $|\psi_{\text{inc}} + \psi_{\text{sca},0}|$ b) RCS

Figure 3.3. *a) Modulus of the total field versus the abscissa x and the height z. b) RCS in dBm scale versus the observation angle θ_{sca}. The parameters are the same as in Figure 3.2*

As expected, a coupling between the two cylinders is observed. Figure 3.3b) shows clearly that the scattered field is strongly modified between the two cylinders.

For the TE and TM polarizations, Figures 3.4 and 3.5 show the same variation as in Figure 3.3, respectively, but the cylinders are PC. In comparison to Figure 3.3, the coupling between the two cylinders is stronger because no field is absorbed inside the cylinders, unlike dielectric cylinders.

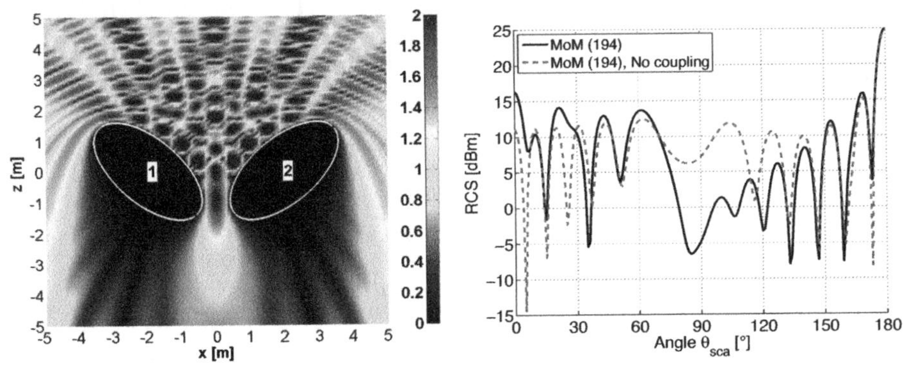

a) $|\psi_{\text{inc}} + \psi_{\text{sca},0}|$ b) RCS

Figure 3.4. *Same variation as in Figure 3.3, but the cylinders are perfectly conducting*

3.2.3. *Method of moments for P scatterers*

From equation [3.12], it is easy to generalize the MoM for P scatterers. Then, we have:

$$\bar{\boldsymbol{Z}} = \begin{bmatrix} \bar{\boldsymbol{Z}}_1 & \bar{\boldsymbol{Z}}_{12} & \ldots \bar{\boldsymbol{Z}}_{1P} \\ \bar{\boldsymbol{Z}}_{21} & \bar{\boldsymbol{Z}}_2 & \ldots \bar{\boldsymbol{Z}}_{2P} \\ \vdots & \ldots & \ddots \vdots \\ \bar{\boldsymbol{Z}}_{P1} & \bar{\boldsymbol{Z}}_{P2} & \ldots \bar{\boldsymbol{Z}}_P \end{bmatrix}, \boldsymbol{X} = \begin{bmatrix} \boldsymbol{X}_1 \\ \boldsymbol{X} \\ \vdots \\ \boldsymbol{X}_P \end{bmatrix}, \qquad [3.25]$$

where

$$\bar{\boldsymbol{Z}}_{ij} = \begin{bmatrix} \bar{\boldsymbol{A}}_{ij} & \bar{\boldsymbol{B}}_{ij} \\ \bar{\boldsymbol{0}} & \bar{\boldsymbol{0}} \end{bmatrix}, \quad \bar{\boldsymbol{Z}}_i = \begin{bmatrix} \bar{\boldsymbol{A}}_i & \bar{\boldsymbol{B}}_i \\ \bar{\boldsymbol{C}}_i & \dfrac{1}{\rho_{i0}}\bar{\boldsymbol{D}}_i \end{bmatrix}, \qquad [3.26]$$

and integers i and $j \neq i$ range from 1 to P. If the number of samples on scatterer i is N_i, then the size of the resulting impedance matrix is $(2\sum_{i=1}^{P} N_i) \times (2\sum_{i=1}^{P} N_i)$, which significantly increases the number of unknowns of the scatterers equaling $2\sum_{i=1}^{P} N_i$ (currents ψ_i and $\partial\psi_i/\partial n$ discretized on each scatterer).

In addition, the complexity of programming also increases significantly. Nevertheless, if the scatterers are assumed to be PC, then for the TE and TM polarizations, $\{\bar{\boldsymbol{Z}}_i = \bar{\boldsymbol{B}}_i, \bar{\boldsymbol{Z}}_{ij} = \bar{\boldsymbol{B}}_{ij}\}$ and $\{\bar{\boldsymbol{Z}}_i = \bar{\boldsymbol{A}}_i, \bar{\boldsymbol{Z}}_{ij} = \bar{\boldsymbol{A}}_{ij}\}$, respectively. In addition, the size of the impedance matrix is reduced to $\sum_{i=1}^{P} N_i \times \sum_{i=1}^{P} N_i$.

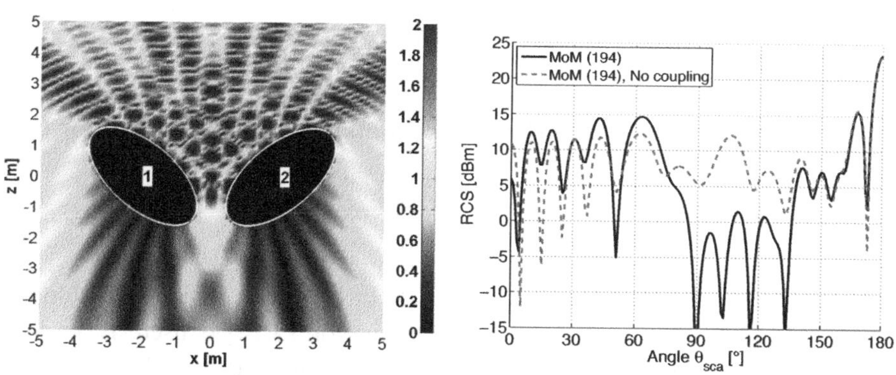

a) $|\psi_{\text{inc}} + \psi_{\text{sca},0}|$ b) RCS

Figure 3.5. *Same variation as in Figure 3.4, but for the TM polarization*

For the TE and TM polarizations, Figure 3.6 shows the total field scattered by six scatterers (three PC elliptical cylinders and three PC plates).

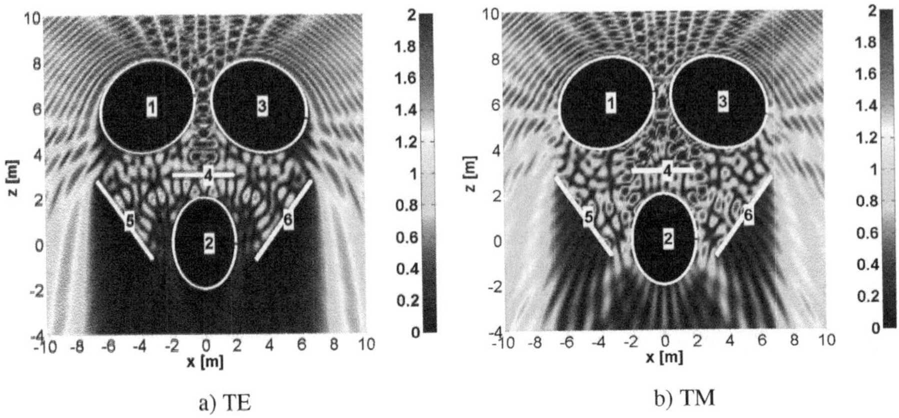

Figure 3.6. *Modulus of the total field* $\psi_{inc} + \psi_{sca,0}$ *versus the abscissa* x *and the height* z. *a) TE polarization. b) TM polarization. The three elliptical PC cylinders have semi-major* $\{a_1 = a_3 = 3, a_2 = 2\}\lambda_0$ *and semi-minor* $\{b_1 = b_2 = b_3 = 2\}\lambda_0$ *axes, their centers are* $C_1 = (-3.5, 6)\lambda_0$, $C_2 = (0, 0)$ *and* $C_3 = (+3.5, 6)\lambda_0$, *and their rotation angles are* $\alpha_1 = 10^o$, $\alpha_2 = 0$ *and* $\alpha_3 = -10^o$. *The lengths of the three PC plates are* $L_4 = 4\lambda_0$ *and* $L_5 = L_6 = 5\lambda_0$, *their centers are* $C_4 = (0, 3)\lambda_0$, $C_5 = (-5, 1)\lambda_0$ *and* $C_6 = (+5, 1)\lambda_0$ *and their rotation angles are* $\alpha_4 = 0$, $\alpha_5 = -\alpha_6 = -\pi/4$ ($\lambda_0 = 1$ *m)*

3.3. Efficient inversion of the impedance matrix: E-PILE method for two scatterers

3.3.1. *Mathematical formulation*

To efficiently solve the linear system $\bar{Z}X = b$, a decomposition by blocks of the the impedance matrix expressed by equation [3.12] is used. From this decomposition, the inversion of $\bar{Z}$ can also be expressed by submatrices as [PRE 92]:

$$\bar{Z}^{-1} = \begin{bmatrix} \bar{T} & \bar{U} \\ \bar{V} & \bar{W} \end{bmatrix}, \qquad [3.27]$$

where

$$\begin{cases} \bar{T} = \left(\bar{Z}_1 - \bar{Z}_{21}\bar{Z}_2^{-1}\bar{Z}_{12}\right)^{-1} \\ \bar{U} = -\left(\bar{Z}_1 - \bar{Z}_{21}\bar{Z}_2^{-1}\bar{Z}_{12}\right)^{-1}\bar{Z}_{21}\bar{Z}_2^{-1} \\ \bar{V} = -\bar{Z}_2^{-1}\bar{Z}_{12}\left(\bar{Z}_1 - \bar{Z}_{21}\bar{Z}_2^{-1}\bar{Z}_{12}\right)^{-1} \\ \bar{W} = \bar{Z}_2^{-1} + \bar{Z}_2^{-1}\bar{Z}_{12}\left(\bar{Z}_1 - \bar{Z}_{21}\bar{Z}_2^{-1}\bar{Z}_{12}\right)^{-1}\bar{Z}_{21}\bar{Z}_2^{-1} \end{cases}, \qquad [3.28]$$

and the unknown vector $\boldsymbol{X}$ is obtained as follows:

$$\begin{bmatrix} \boldsymbol{X}_1 \\ \boldsymbol{X}_2 \end{bmatrix} = \bar{Z}^{-1}\begin{bmatrix} \boldsymbol{b}_1 \\ \boldsymbol{b}_2 \end{bmatrix} = \begin{bmatrix} \bar{T}\boldsymbol{b}_1 + \bar{U}\boldsymbol{b}_2 \\ \bar{V}\boldsymbol{b}_1 + \bar{W}\boldsymbol{b}_2 \end{bmatrix}. \qquad [3.29]$$

By using equations [3.28] and [3.29], the currents on scatterer 1, $\boldsymbol{X}_1$, can be expressed as:

$$\boldsymbol{X}_1 = \left(\bar{Z}_1 - \bar{Z}_{21}\bar{Z}_2^{-1}\bar{Z}_{12}\right)^{-1}\boldsymbol{b}_1 - \left(\bar{Z}_1 - \bar{Z}_{21}\bar{Z}_2^{-1}\bar{Z}_{12}\right)^{-1}\bar{Z}_{21}\bar{Z}_2^{-1}\boldsymbol{b}_2, \qquad [3.30]$$

which leads to:

$$\begin{aligned} \boldsymbol{X}_1 &= \left(\bar{Z}_1 - \bar{Z}_{21}\bar{Z}_2^{-1}\bar{Z}_{12}\right)^{-1}\left(\boldsymbol{b}_1 - \bar{Z}_{21}\bar{Z}_2^{-1}\boldsymbol{b}_2\right) \\ &= \left(\bar{I} - \bar{Z}_1^{-1}\bar{Z}_{21}\bar{Z}_2^{-1}\bar{Z}_{12}\right)^{-1}\bar{Z}_1^{-1}\left(\boldsymbol{b}_1 - \bar{Z}_{21}\bar{Z}_2^{-1}\boldsymbol{b}_2\right), \end{aligned} \qquad [3.31]$$

where $\bar{I}$ is the identity matrix. Let us introduce the characteristic matrix $\bar{M}_{c,1}$ as:

$$\bar{M}_{c,1} = \bar{Z}_1^{-1}\bar{Z}_{21}\bar{Z}_2^{-1}\bar{Z}_{12}. \qquad [3.32]$$

The first term in equation [3.31] can be expanded as an infinite series (Taylor series expansion of the inverse of the Shur complement) over p:

$$\left(\bar{I} - \bar{Z}_1^{-1}\bar{Z}_{21}\bar{Z}_2^{-1}\bar{Z}_{12}\right)^{-1} = \left(\bar{I} - \bar{M}_{c,1}\right)^{-1} = \sum_{p=0}^{p=\infty}\bar{M}_{c,1}^{p}. \qquad [3.33]$$

For the numerical computation, the sum must be lowered at the order $P_{\text{E-PILE}}$. From equations [3.31] and [3.33], the currents on scatterer 1, $\boldsymbol{X}_1$, are then expressed as:

$$\boldsymbol{X}_1 = \left[\sum_{p=0}^{p=P_{\text{E-PILE}}} \bar{\boldsymbol{M}}_{c,1}^{p}\right] \bar{\boldsymbol{Z}}_1^{-1}\left(\boldsymbol{b}_1 - \bar{\boldsymbol{Z}}_{21}\bar{\boldsymbol{Z}}_2^{-1}\boldsymbol{b}_2\right) = \sum_{p=0}^{p=P_{\text{E-PILE}}} \boldsymbol{Y}_1^{(p)}, \quad [3.34]$$

where

$$\begin{cases} \boldsymbol{Y}_1^{(0)} = \bar{\boldsymbol{Z}}_1^{-1}\left(\boldsymbol{b}_1 - \bar{\boldsymbol{Z}}_{21}\bar{\boldsymbol{Z}}_2^{-1}\boldsymbol{b}_2\right) & \text{for } p = 0 \\ \boldsymbol{Y}_1^{(p)} = \bar{\boldsymbol{M}}_{c,1}\boldsymbol{Y}_1^{(p-1)} & \text{for } p > 0 \end{cases}. \quad [3.35]$$

The currents on scatterer 2, $\boldsymbol{X}_2$, are obtained from equations [3.34], [3.35] and [3.32] by substituting, the subscripts $\{2, 1, 21, 12\}$ for the subscripts $\{1, 2, 12, 21\}$, respectively.

We define the norm $\|\bar{\boldsymbol{M}}_{c,1}\|$ of a complex matrix by its spectral radius, that is the modulus of its eingenvalue, which has the highest modulus. Expansion [3.33] is then valid if $\|\bar{\boldsymbol{M}}_{c,1}\|$ is strictly smaller than 1.

The physical interpretation of $\bar{\boldsymbol{M}}_{c,1}$ is shown in Figure 3.7: in the zeroth-order term, $\bar{\boldsymbol{Z}}_1^{-1}$ accounts for the local interactions on scatterer 1, so $\boldsymbol{Y}_1^{(0)}$ corresponds to the contribution of the scattering on scatterer 1 when it is illuminated by the direct incident field ($\boldsymbol{b}_1$) and the direct scattered field by scatterer 2 ($-\bar{\boldsymbol{Z}}_{21}\bar{\boldsymbol{Z}}_2^{-1}\boldsymbol{b}_2$). Indeed, $\bar{\boldsymbol{Z}}_2^{-1}$ accounts for the local interactions on scatterer 2, and $\bar{\boldsymbol{Z}}_{21}$ propagates the field on scatterer 2 toward scatterer 1. In the first-order term, $\boldsymbol{Y}_1^{(1)} = \bar{\boldsymbol{M}}_{c,1}\boldsymbol{Y}_1^{(0)}$, $\bar{\boldsymbol{Z}}_{12}$ propagates the resulting field information, $\boldsymbol{Y}_1^{(0)}$, toward scatterer 2, $\bar{\boldsymbol{Z}}_2^{-1}$ accounts for the local interactions on this scatterer and $\bar{\boldsymbol{Z}}_{21}$ re-propagates the resulting field information toward scatterer 1; finally, $\bar{\boldsymbol{Z}}_1^{-1}$ updates the field values on scatterer 1. So the characteristic matrix $\bar{\boldsymbol{M}}_{c,1}$ is related to a back-and-forth between the two scatterers. In conclusion, the order $P_{\text{E-PILE}}$ of PILE corresponds to the number of back-and-forths between scatterers 1 and 2. The same conclusion holds for $\bar{\boldsymbol{M}}_{c,2}$.

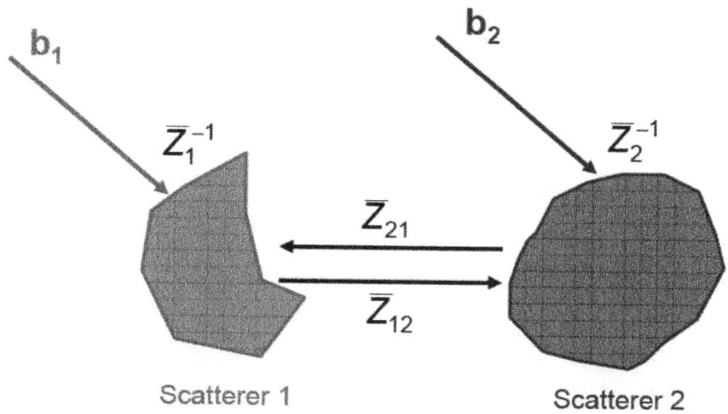

Figure 3.7. *Illustration of the physical interpretation of the E-PILE method*

3.3.2. *Numerical results*

3.3.2.1. *Two perfectly conducting plates*

To confirm that the E-PILE order is related to the number of reflections between the two scatterers, we will choose scenarios for which this phenomenon is clearly identified.

Figure 3.8 shows the modulus of the total field versus the abscissa x in meters and the height z in meters. Two identical PC plates are considered. The simulations parameters are $\lambda_0 = 1$ m, $\theta_{\text{inc}} = 0$, $\psi_{\text{inc},0} = 1$, the lengths of the plates are $L_1 = L_2 = 5\lambda_0$, their rotation angles are $\alpha_1 = -\alpha_2 = \pi/4$, their centers have coordinates $C_{1,2} = (\mp 2.5, 0)\lambda_0$ and the polarization is TE. Figure 3.9 shows the corresponding RCS in dBm scale versus the scattering angle θ_{sca}. In the figure, for a given E-PILE order, the number in parentheses is the residual relative error defined as:

$$\epsilon_{\text{E-PILE}} = \frac{\text{norm}_{\theta_{\text{sca}}}(\text{RCS}_{\text{E-PILE}} - \text{RCS}_{\text{LU}})}{\text{norm}_{\theta_{\text{sca}}}(\text{RCS}_{\text{LU}})}, \qquad [3.36]$$

where the symbol "norm" stands for the norm 2. RCS_{LU} stands for the RCS that is computed from a direct LU inversion of the impedance matrix of the two scatterers and, $\text{RCS}_{\text{E-PILE}}$ that computed from E-PILE.

Figure 3.9 shows that the coupling must be taken into account and the results match well with those computed from LU for $P_{\text{E-PILE}} = 0$. As shown in Figure 3.8, this corresponds to one reflection between the two plates. For this scenario and from a geometrical point of view, only one reflection between the two plates occurs, and for $\theta_{\rm inc} = 0$, the RCS is then at a maximum for a scattering angle $\theta_{\rm sca} = 0$.

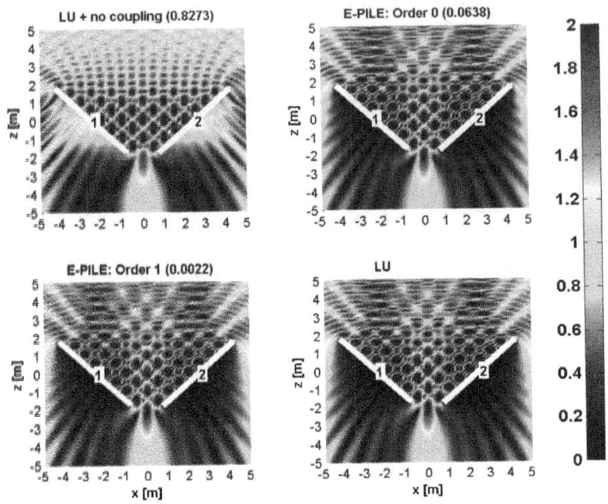

Figure 3.8. *Modulus of the total field versus the abscissa x in meters and the height z in meters. Two identical perfectly conducting plates are considered. $\lambda_0 = 1$ m, $\theta_{inc} = 0$, $\psi_{inc,0} = 1$; the lengths of the plates are $5\lambda_0$, their rotation angles are $\alpha_1 = -\alpha_2 = \pi/4$ (angles defined in counterclockwise), their centers have coordinates $C_{1,2} = (\mp 2.5, 0)\lambda_0$, and the polarization is TE*

Figure 3.10 shows the same results as in Figure 3.8, but $L_1 = L_2 = 6\lambda_0$, $\alpha_1 = \alpha_2 = 0$ and $C_{1,2} = (0, \mp 2.5)\lambda_0$. Figure 3.11 shows the corresponding RCS in dBm scale versus the scattering angle $\theta_{\rm sca}$.

For this example, from a geometrical point of view, the number of reflections inside the opened cavity (space between the two plates) is given by the integer part of $L/(d \tan\theta_{\rm inc})$, giving 2 for $d = 5\lambda_0$. It is consistent with Figures 3.10 and 3.9, where the convergence is achieved for $P_{\text{E-PILE}} = 2$ (for a given precision).

To know the convergence rate of E-PILE, we therefore need to calculate the norm of the characteristic matrix, $||\bar{\boldsymbol{M}}_c||$, which equals the modulus of its eingenvalue, which has the highest modulus. For Figures 3.9 and 3.11, $||\bar{\boldsymbol{M}}_c|| = \{0.036, 0.956\}$, respectively. For a large problem, the computation of this norm is very time consuming and it is not recommended.

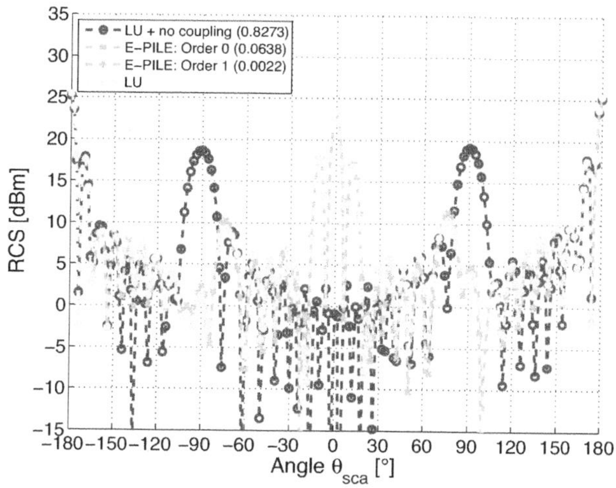

Figure 3.9. *RCS in dBm scale versus the scattering angle θ_{sca}. The parameters are the same as in Figure 3.8*

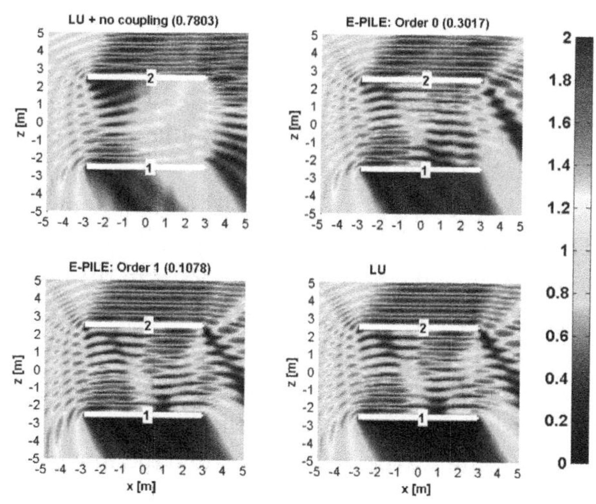

Figure 3.10. *Same results as in Figure 3.8, but $\theta_{inc} = 30^o$, $L_1 = L_2 = 6\lambda_0$, $\alpha_1 = \alpha_2 = 0$ and $C_{1,2} = (0, \mp 2.5)\lambda_0$*

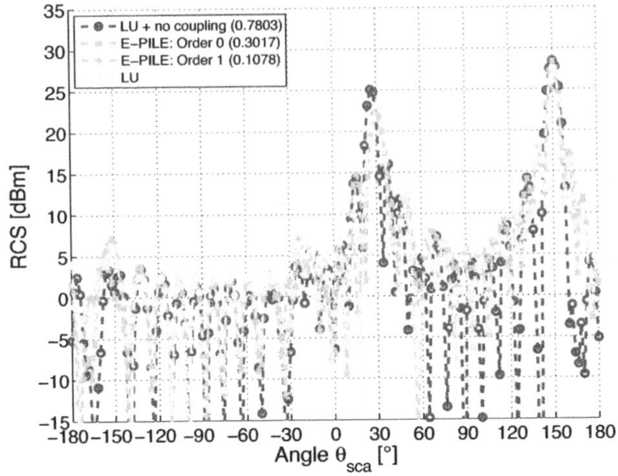

Figure 3.11. *RCS in dBm scale versus the scattering angle θ_{sca}. The parameters are the same as in Figure 3.10*

3.3.2.2. *Two elliptical perfectly conducting cylinders*

Figure 3.12 shows the modulus of the total field in dB scale versus the abscissa x in m and the height z in meters. The simulation parameters are $\lambda_0 = 1$ meters, $a_1 = a_2 = \lambda_0$ (semi-major axis), $b_1 = b_2 = 3\lambda_0/2$ (semi-minor axis), $C_{1,2} = (0, \mp 2)\lambda_0$ (centers), $\alpha_{1,2} = \pm 30^{\text{o}}$ (rotation angles), $\psi_{\text{inc},0} = 1$ and $\theta_{\text{inc}} = 0$, TE polarization and the two elliptical cylinders are assumed to be PC. Figure 3.13 shows the corresponding RCS in dBm scale versus the scattering angle θ_{sca}. As the order $P_{\text{E-PILE}}$ increases, Figure 3.12 shows that the field computed from E-PILE between the two cylinders and behind the upper cylinder converges toward that computed from LU. Between the two cylinders, the fields do not vanish because the first cylinder is excited by the second cylinder due to the creeping waves, whereas behind the first cylinder, the total field vanishes in the shadow zone. Figure 3.13 shows that the convergence is achieved for $P_{\text{E-PILE}} = 3$.

For a scenario made up of an object (like an elliptical cylinder, a plate and a cross) above a rough surface of sea-like and Gaussian spectra, Kubické *et al.* [KUB 08, KUB 10c, KUB 10b] showed that the E-PILE method always converges.

For Figure 3.13, $\|\bar{M}_c\| = 0.221$.

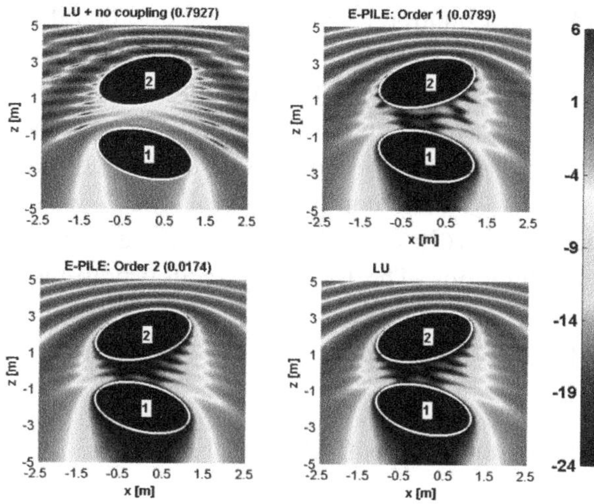

Figure 3.12. *Modulus of the total field in dB scale versus the abscissa x in meters and the height z in meters. $\lambda_0 = 1$ m, $a_1 = a_2 = \lambda_0$ (semi-major axis), $b_1 = b_2 = 3\lambda_0/2$ (semi-minor axis), $C_{1,2} = (0, \mp 2)\lambda_0$ (centers), $\alpha_{1,2} = \pm 30^o$ (rotation angles), $\psi_{inc,0} = 1$ and $\theta_{inc} = 0$, TE polarization and the two elliptical cylinders are assumed to be perfectly conducting*

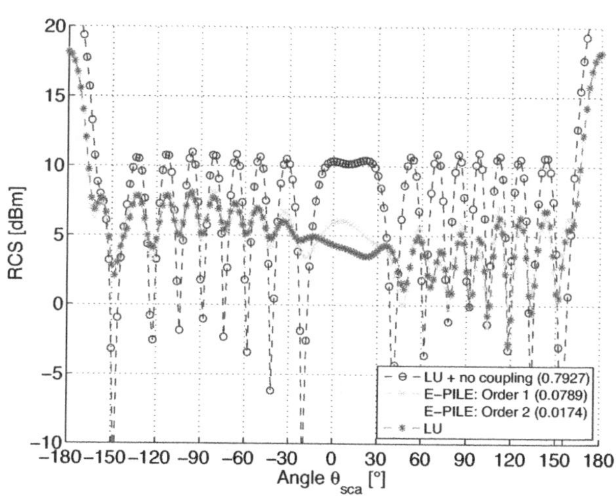

Figure 3.13. *RCS in dBm scale versus the scattering angle θ_{sca}. The parameters are the same as in Figure 3.12*

In comparison to an LU inversion, one of the advantages of the E-PILE method is its ability to quantify the coupling between the two scatterers within the order $P_{\text{E-PILE}}$. Equation [3.35] clearly shows that the calculation of $\bar{\boldsymbol{Z}}_i^{-1}\boldsymbol{u}$ is required, in which $\bar{\boldsymbol{Z}}_i$ is the impedance matrix of scatterer i in free space. Another advantage of the E-PILE method is its ability to calculate this matrix-vector product by fast numerical methods that already exist for a single scatterer (in free space). In comparison to a direct LU inversion of complexity $\mathcal{O}((N_1 + N_2)^3)$, the complexity of E-PILE, which is similar to LU without hybridization ($\mathcal{O}(N_1^3 + N_2^3)$), can be significantly reduced by combining such accelerations. The following section presents how the local interactions (related to $\bar{\boldsymbol{Z}}_i^{-1}\boldsymbol{u}$) on each scatterer can be computed by the FB and the PO in the E-PILE algorithm.

3.4. E-PILE method combined with PO and FB

This section presents how the PO and FB methods, applied to calculate the local interactions on a single scatterer, can be hybridized in the E-PILE algorithm to compute the field scattered by two scatterers.

3.4.1. *E-PILE hybridized with FB*

Equation [3.35] clearly shows that the calculation of $\bar{\boldsymbol{Z}}_i^{-1}\boldsymbol{u}$ is required, in which $\bar{\boldsymbol{Z}}_i$ is the impedance matrix of scatterer i in free space. To calculate the local interactions on a rough surface, the FB method can be applied. For any scatterer above a rough surface, in [KUB 08], [KUB 10c] and [KUB 10b], it was shown that the order of convergence P_{FB} is obtained by considering only the scattering from the single rough surface (without the scatterer). Physically, this conclusion can be explained by the fact that the inversion of the impedance matrix is independent of the incident field $\boldsymbol{u}$.

Thus, if we consider the rough surface (scatterer 1) giving the NRCS plotted in Figure 2.28a) ($\lambda_0 = 1$ m, $N_{\lambda_0} = 10$, $L = 300\lambda_0$, $\sigma_z = 0.5\lambda_0$, $L_c = 5\lambda_0$, $\theta_{\text{inc}} = 30°$, the surface height autocorrelation function is Gaussian and the Dirichlet boundary condition), $P_{\text{FB}} = 5$. A circular cylinder (scatterer 2) of radius $a_1 = 2\lambda_0$ and center $C_1 = (0, 3\lambda_0)$ is located above the surface. The scene is shown in Figure 3.14a) and the corresponding NRCS is plotted in Figure 3.14b).

In the figure, for a given E-PILE order combined with a fifth-order FB, the number in parentheses is the residual relative error defined as:

$$\epsilon_{\text{E-PILE+FB}} = \frac{\text{norm}_{\theta_{\text{sca}}}\left(\text{NRCS}_{\text{E-PILE+FB}} - \text{NRCS}_{\text{LU}}\right)}{\text{norm}_{\theta_{\text{sca}}}\left(\text{NRCS}_{\text{LU}}\right)}, \qquad [3.37]$$

where the symbol "norm" stands for the norm 2.

As the E-PILE order increases, Figure 3.14b) shows that the results converge toward those obtained from an LU inversion and the residual error at $P_{\text{E-PILE}} = 8$ is similar to that obtained when the surface is alone (without the object, see Figure 2.28a)). In addition, E-PILE+FB converges rapidly, and the convergence order is $P_{\text{E-PILE}} = 8$ (for a given precision). It depends on the expected precision.

Figure 3.15 shows the same results as in Figure 3.14, but the object is a plate of length $L_1 = 10\lambda_0$, of center $C_1 = (0, 5\lambda_0)$, of rotation angle $\alpha_1 = 0$ and $N_{\lambda_0} = 10$. Figure 3.15b) shows that E-PIPLE+FB converges rapidly ($P_{\text{E-PILE}} = 1$) and like previously, the residual error is not zero because the FB is applied to calculate the local interactions on the rough surface.

The complexity of E-PILE+FB is $\mathcal{O}(N_1N_2 + N_1^2 + N_2^3)$ instead of $\mathcal{O}((N_1 + N_2)^3)$ if a direct LU inversion is used. $\mathcal{O}(N_1N_2)$ corresponds to the complexity of the vector-matrix products, $\mathcal{O}(N_2^2)$ corresponds to the complexity of FB applied to compute the local interactions on the rough surface (scatterer 2) and $\mathcal{O}(N_1^3)$ corresponds to the complexity of LU to compute the local interactions on the object (scatterer 1). Combining the SA [CHO 02, CHO 00, TOR 00, TOR 02] with FB, the complexity $\mathcal{O}(N_2^2)$ becomes $\mathcal{O}(N_2)$ and then the E-PILE+FB-SA method is very efficient for calculating the field scattered by an object above a very large rough surface (in comparison to the object) [KUB 10a, KUB 10c, KUB 10b]. In addition, with the SA, the memory space requirement decreases from $\mathcal{O}(N_2^2)$ (for the FB) to $\mathcal{O}(N_2)$, because only the elements of the impedance matrix of the strong interactions must be stored.

Another means of decreasing the memory space requirement and the complexity is to hybridize E-PILE with asymptotic methods such as PO. This is described in the next section.

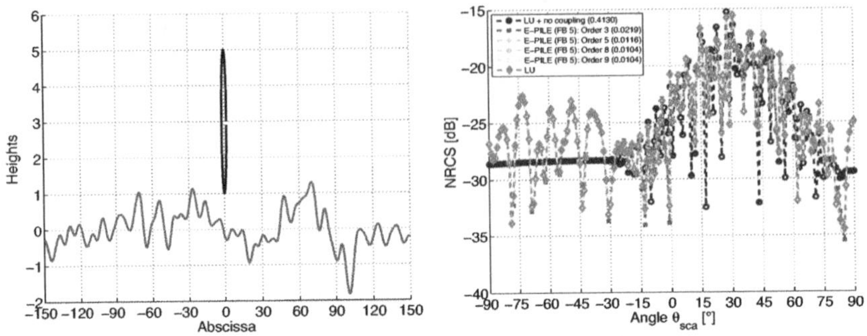

a) Scene b) NRCS

Figure 3.14. *NRCS in dB scale versus the observation angle θ_{sca} computed from LU and E-PILE combined with FB. $\theta_{inc} = 30^o$ and the Neumann boundary condition (TM polarization). Rough surface: $\lambda_0 = 1$ m, $N_{\lambda_0} = 10$, $L_1 = 300\lambda_0$, $\sigma_z = 0.5\lambda_0$, $L_c = 5\lambda_0$, and the surface height autocorrelation function is Gaussian. Circular cylinder: $N_{\lambda_0} = 10$, $a_1 = 2\lambda_0$ and center $C_1 = (0, 3\lambda_0)$*

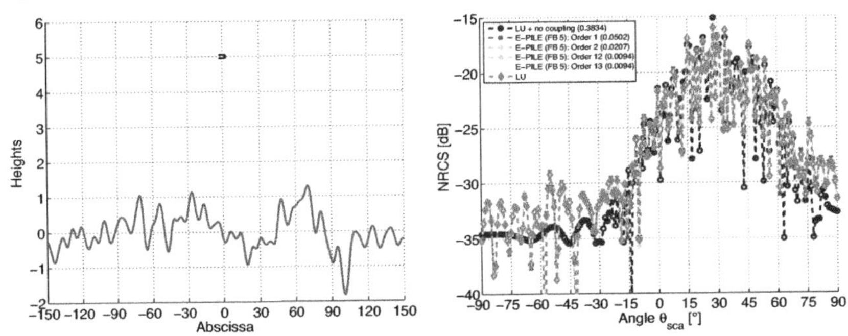

a) Scene b) NRCS

Figure 3.15. *Same results as in Figure 3.14 but the object is a plate of length $L_1 = 10\lambda_0$, of center $C_1 = (0, 5\lambda_0)$, of rotation angle $\alpha_2 = 0$ and $N_{\lambda_0} = 10$*

3.4.2. *E-PILE hybridized with PO*

This section focuses on the integration of PO in the E-PILE algorithm. We consider only the PO at the first order, meaning that the multiple reflections on the same scatterer are neglected. As shown in [KUB 10a], it is not difficult to include the higher order reflections but the complexity of programming increases.

3.4.2.1. *The Neumann boundary condition*

For the TM polarization and for a PC object (scatterer 1) in free space, the total field on the object surface due to a single reflection is given with PO approximation by equation [2.62]:

$$\psi_1(\boldsymbol{r}) = 2\psi_{\text{inc}}(\boldsymbol{r}) \quad \forall \boldsymbol{r} \in S_{\text{Ill}}, \tag{3.38}$$

where S_{Ill} is the illuminated surface on the object excited by the incident field $\psi_{\text{inc}}(\boldsymbol{r})$, and $\psi(\boldsymbol{r})$ is the total field on the object.

If we consider two different sources that illuminate the object, we can write under PO approximation:

$$\psi_1(\boldsymbol{r}) = 2\left[\,\Xi_1(\boldsymbol{r})\psi_{\text{inc}}(\boldsymbol{r}) + \Xi_{21}(\boldsymbol{r})\psi_{21}(\boldsymbol{r})\right] = 2\psi'_{\text{inc}}(\boldsymbol{r}), \tag{3.39}$$

where Ξ_1 and Ξ_{21} are two shadowing functions (Boolean) defined as:

$$\Xi_1(\boldsymbol{r}) = \begin{cases} 1 \; \forall \boldsymbol{r} \in S_{\text{Ill}} \\ 0 \text{ otherwise} \end{cases}, \tag{3.40}$$

and

$$\Xi_{21}(\boldsymbol{r}) = \begin{cases} 1 \; \forall \boldsymbol{r} \in S_{\text{Ill}-21} \\ 0 \text{ otherwise} \end{cases}, \tag{3.41}$$

where $S_{\text{Ill}-21}$ is the surface of the object illuminated by the field $\psi_{21}(\boldsymbol{r})$.

By sampling the object surface into N_1 elements, equation [3.39] can be converted into

$$\boldsymbol{X}_1^{\text{PO}} = 2\left(\bar{\Xi}_1\boldsymbol{b}_1 + \bar{\Xi}_{21}\boldsymbol{b}_{21}\right) = 2\boldsymbol{u}', \tag{3.42}$$

where $\boldsymbol{X}_1^{\text{PO}}$ is the unknown vector whose elements are the total field on the object $\psi(\boldsymbol{r}_{1,n})$ sampled on the surface ($n \in [1; N_1]$ and $\boldsymbol{r}_{1,n} \in \Xi_1$). $\bar{\Xi}_1$ and $\bar{\Xi}_{21}$ are diagonal matrices, in which some elements are set to zero due to the shadowing (sampling of the shadowing functions $\Xi_1(\boldsymbol{r}_{1,n})$ and $\Xi_{21}(\boldsymbol{r}_{1,n})$). Their elements are defined as:

$$\Xi_{1,nn} = \begin{cases} 1 \ \forall \boldsymbol{r}_{1,n} \in S_{\text{III}} \\ 0 \text{ otherwise} \end{cases}, \qquad [3.43]$$

and

$$\Xi_{21,nn} = \begin{cases} 1 \ \forall \boldsymbol{r}_{1,n} \in S_{\text{III}-21} \\ 0 \text{ otherwise} \end{cases}. \qquad [3.44]$$

By considering the zeroth order ($P_{\text{E-PILE}} = 0$) of E-PILE, equation [3.42] is quite similar to [3.35]. Indeed, the unknowns $\boldsymbol{X}_1^{(0)}$ on the object can be expressed as:

$$\boldsymbol{X}_1^{(0)} = \bar{\boldsymbol{Z}}_1^{-1} \boldsymbol{u}, \qquad [3.45]$$

where $\boldsymbol{u}$ can be seen as an incident field on the object defined as:

$$\boldsymbol{u} = \boldsymbol{b}_1 - \bar{\boldsymbol{Z}}_{21} \bar{\boldsymbol{Z}}_2^{-1} \boldsymbol{b}_2. \qquad [3.46]$$

In the E-PILE algorithm, the excitation can be split up into two sources: one related to the direct illumination of the incident field $\boldsymbol{b}_1$, and the other one related to the coupling (multiple interactions) with the second scatterer $\bar{\boldsymbol{Z}}_{21}\boldsymbol{v}$, where $\boldsymbol{v}$ are the currents on scatterer 2. Thus, the excited surfaces on the object are not the same according to the source. For example, for a plate (scatterer 1) above a rough surface (scatterer 2), the incident field $\boldsymbol{b}_1$ illuminates the top side of the plate, whereas the field scattered from the rough surface toward the object illuminates the bottom side of the plate.

Thus, from equations [3.46], [3.45] and [3.42], we have:

$$\boldsymbol{X}_1^{(0)} = 2 \underbrace{\left(\boldsymbol{b}_1' - \bar{\boldsymbol{Z}}_{21}' \bar{\boldsymbol{Z}}_2^{-1} \boldsymbol{b}_2 \right)}_{\boldsymbol{u}} = 2\boldsymbol{u}', \qquad [3.47]$$

where the elements of $\bar{\boldsymbol{Z}}_1^{-1}$ are $Z_{1,mn} = 2\delta_{m,n}$ (Kronecker symbol defined as $\delta_{m,n} = 1$ if $m = n$, 0 otherwise), $\boldsymbol{b}_1' = \bar{\Xi}_1 \boldsymbol{b}_1$ and $\bar{\boldsymbol{Z}}_{21}' = \bar{\Xi}_{21} \bar{\boldsymbol{Z}}_{21}$. Thus, the inversion of $\bar{\boldsymbol{Z}}_1$ becomes analytic and straightforward. The shadowing is taken into account within the source vector $\boldsymbol{u}'$. Moreover, the matrix-vector product $\bar{\boldsymbol{Z}}_1^{-1}\boldsymbol{u}$ becomes a scalar-vector product $2\boldsymbol{u}'$.

At the order $P_{\text{E-PILE}}$, since $\bar{\boldsymbol{M}}_{c,1}$ is related to a back-and-forth between the two scatterers (which rigorously takes into account the coupling effect), the total field $\boldsymbol{X}_1$ on scatterer 1 is expressed from [3.34], in which $\boldsymbol{X}_1^{(0)}$ is given by [3.47] and the p-th order is given by:

$$\boldsymbol{X}_1^{(p)} = \bar{\boldsymbol{M}}'_{c,1}\boldsymbol{X}_1^{(p-1)}, \qquad [3.48]$$

where $\bar{\boldsymbol{M}}'_{c,1} = 2\bar{\boldsymbol{Z}}'_{21}\bar{\boldsymbol{Z}}_2^{-1}\bar{\boldsymbol{Z}}_{12}$.

The elements of the vector $\boldsymbol{b}'_1 = \bar{\bar{\Xi}}_1\boldsymbol{b}_1$ and of the matrix $\bar{\boldsymbol{Z}}'_{21} = \bar{\bar{\Xi}}_{21}\bar{\boldsymbol{Z}}_{21}$ are defined from geometrical considerations as:

$$b'_{1,n} = b_{1,n}\frac{1 - \operatorname{sgn}\left[\hat{\boldsymbol{n}}_{1,n}\cdot\hat{\boldsymbol{k}}_{\text{inc}}\right]}{2} = b_{1,n}\frac{1 + \operatorname{sgn}\left[v_{1,n}\gamma_{1,n}\sin\theta_{\text{inc}} + v_{1,n}\cos\theta_{\text{inc}}\right]}{2}, \qquad [3.49]$$

and

$$\begin{aligned} Z'_{21,mn} &= Z_{21,mn}\frac{1 - \operatorname{sgn}\left[(\boldsymbol{r}_{1,m} - \boldsymbol{r}_{2,n})\cdot\hat{\boldsymbol{n}}_{1,m}\right]}{2} \\ &= Z_{21,mn}\frac{1 + \operatorname{sgn}\left[(x_{1,m} - x_{2,n})v_{1,m}\gamma_{1,m} - (z_{1,m} - z_{2,n})v_{1,m}\right]}{2}, \end{aligned} \qquad [3.50]$$

where $v_{1,m}$ gives the sign of the vector $\hat{\boldsymbol{n}}_1$ normal to scatterer 1 ($v_1 = \operatorname{sgn}(\hat{\boldsymbol{n}}_1\cdot\hat{\boldsymbol{z}})$), and $\gamma_{1,m}$ is the slope of scatterer 1 at the point $\boldsymbol{r}_{1,m}$.

In conclusion, the impedance matrix $\bar{\boldsymbol{Z}}_1$ is not stored ($\mathcal{O}(N_1^2) \to 1$), its inversion is analytic (not computed: $\mathcal{O}(N_1^3) \to 1$) and the computing time is also reduced for the matrix-vector product $\bar{\boldsymbol{Z}}_1^{-1}\boldsymbol{v}$ (for any incident field vector $\boldsymbol{v}$); PO shadowing effects being geometrically taken into account during the incident vector and coupling matrix fillings.

The PO approximation can also be applied onto scatterer 2 by the substitution of subscripts $\{1, 2, 21, 12\}$ for $\{2, 1, 12, 21\}$. The resulting complexity of the E-PILE method, named as E-PILE + PO (scatterer 1) + PO (scatterer 2), is then $\mathcal{O}(N_1N_2)$, corresponding to the matrix-vector products complexity, instead of $\mathcal{O}(N_1^3 + N_2^3 + N_1 N_2) \approx \mathcal{O}(N_1^3 + N_2^3)$ if E-PILE is applied without acceleration.

3.4.2.2. *The Dirichlet boundary condition*

For TE polarization, and for a PC scatterer, the Dirichlet boundary condition implies $\psi(\boldsymbol{r}) = 0$ on the scatterer. Then, $\boldsymbol{X}_1$ is the unknown vector containing the normal derivative of the total field on scatterer 1: $\partial\psi_1(\boldsymbol{r})/\partial n$ sampled on the surface. From the PO approximation, for a scatterer in free space, we have from equation [2.62]:

$$\frac{\partial\psi_1(\boldsymbol{r})}{\partial n} = 2\frac{\partial\psi_{\text{inc}}(\boldsymbol{r})}{\partial n} \quad \forall \boldsymbol{r} \in S_{\text{III}}, \qquad [3.51]$$

and for two sources:

$$\frac{\partial\psi_1(\boldsymbol{r})}{\partial n} = 2\left[\Xi_1(\boldsymbol{r})\frac{\partial\psi_{\text{inc}}(\boldsymbol{r})}{\partial n} + \Xi_{21}(\boldsymbol{r})\frac{\partial\psi_{21}(\boldsymbol{r})}{\partial n}\right] = 2\frac{\partial\psi'_{\text{inc}}(\boldsymbol{r})}{\partial n}, \qquad [3.52]$$

where Ξ_1 and Ξ_{21} are shadowing functions (Boolean) defined from equations [3.40] and [3.41], respectively.

Using the same process as for the TM case, the normal derivative of the total field $\boldsymbol{X}_1$ on scatterer 1 is expressed by equation [3.34], in which $\boldsymbol{X}_1^{(0)}$ is given by [3.47] and the pth order is given by equation [3.48]. In addition, $\bar{\boldsymbol{M}}'_{c,1} = 2\bar{\boldsymbol{Z}}'_{21}\bar{\boldsymbol{Z}}_2^{-1}\bar{\boldsymbol{Z}}_{12}$ with $\boldsymbol{b}'_1 = \bar{\Xi}_1\partial\boldsymbol{b}_1/\partial n_1$ and $\bar{\boldsymbol{Z}}'_{21} = \bar{\Xi}_{21}\partial\bar{\boldsymbol{Z}}_{21}/\partial n_1$. The symbol $\partial\bullet/\partial n_1$ is the normal derivative which operates on the coordinates (x_1, z_1). Like for the TM polarization, the impedance matrix $\bar{\boldsymbol{Z}}_1$ is not stored, its inversion is analytic (not computed) and the computing time is also reduced for the matrix-vector product $\bar{\boldsymbol{Z}}_1^{-1}\boldsymbol{v}$ (for any incident field vector $\boldsymbol{v}$).

The PO approximation can be applied on the scatterer 2 by the substitution of subscripts $\{1, 2, 21, 12\}$ for $\{2, 1, 12, 21\}$ and on scatterer 1, a direct LU inversion is applied. According to the Neumann boundary condition, the MatLab program is more complex because the operator $\partial \bullet /\partial n_1$ must be computed either numerically or analytically.

For the Dirichlet boundary condition, $\bar{\boldsymbol{Z}}_{21} = \bar{\boldsymbol{B}}_{21}$ and from equation [3.17], its elements are:

$$B_{21,mn} = \frac{j|\Delta_{2,n}|\sqrt{1+\gamma_{2,n}^2}}{4}\mathrm{H}_0^{(1)}(k_0\,\|\boldsymbol{r}_{2,n} - \boldsymbol{r}_{1,m}\|). \qquad [3.53]$$

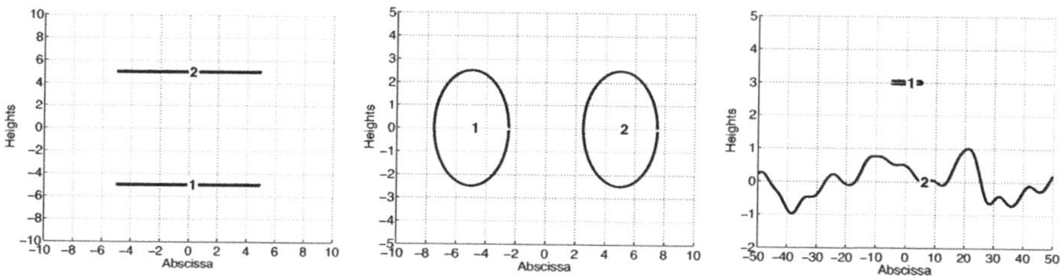

a) Scenario 1 b) Scenario 2 c) Scenario 3

Figure 3.16. *a) Two parallel PC plates with thicknesses ($\lambda_0/20$) of lengths $L_1 = L_2 = 10\lambda_0$, of centers $C_{1,2} = (0, \mp 5\lambda_0)$, of rotation angles $\alpha_{1,2} = 0$ and $\theta_{inc} = 30^o$. b) Two PC circular cylinders of radius $a_1 = a_2 = 2.5\lambda_0$, of centers $C_{1,2} = (\mp 5\lambda_0, 0)$ and $\theta_{inc} = 0$. c) A PC plate above a rough surface, with a plate: $L_1 = 10\lambda_0$, $\alpha_1 = 0$ and $C_1 = (0,3)\lambda_0$; and a rough surface: $L_2 = 300\lambda_0$, $\sigma_z = 0.5\lambda_0$, $L_c = 5\lambda_0$, the surface height autocorrelation function is Gaussian, the parameter of the Thorsos wave (defined from equation* [1.32]*) is $g = L_1/6$. $\theta_{inc} = 30^o$. For the three cases, the polarization is TM and $N_{\lambda_0} = 10$*

Then, the elements of $\partial \bar{\boldsymbol{Z}}_{21}/\partial n_1$ are:

$$\frac{\partial B_{21,mn}}{\partial n_1} = \frac{jk_0 v_{1,m}|\Delta_{2,n}|\sqrt{1+\gamma_{2,n}^2}\,\mathrm{H}_1^{(1)}(k_0 r_{21})}{4r_{21}\sqrt{1+\gamma_{1,m}^2}}\left(z_{21} - \gamma_{1,m} x_{21}\right), \quad [3.54]$$

where $x_{21} = x_{2,n} - x_{1,m}$, $z_{21} = z_{2,n} - z_{1,m}$ and $r_{21} = \sqrt{x_{21}^2 + z_{21}^2}$. In addition, for a plane wave $\psi_{\text{inc}}(\boldsymbol{r}) = \psi_{\text{inc},0} e^{j\boldsymbol{k}_{\text{inc}}\cdot\boldsymbol{r}}$, we have:

$$\frac{\partial \psi_{\text{inc}}(\boldsymbol{r})}{\partial n} = \hat{\boldsymbol{n}} \cdot \boldsymbol{\nabla}\psi_{\text{inc}}(\boldsymbol{r}) = j\hat{\boldsymbol{n}} \cdot \boldsymbol{k}_{\text{inc}}\psi_{\text{inc}}(\boldsymbol{r}). \quad [3.55]$$

3.4.2.3. *Numerical results*

First, we consider the scenario depicted in Figure 3.16a): two parallel PC plates of lengths $L_1 = L_2 = 10\lambda_0$, of centers $C_{1,2} = (0, \mp 5\lambda_0)$, of rotation angles $\alpha_{1,2} = 0$. The incident angle is $\theta_{\text{inc}} = 30^{\text{o}}$, the polarization is TM and $N_{\lambda_0} = 10$.

Figure 3.17 shows the modulus of the currents on scatterer 1 (bottom plate) versus the index n_1. The currents on scatterer 2 (top plate) are not shown. Figure 3.18 shows the RCS versus the scattering angle θ_{sca}. In the figure:

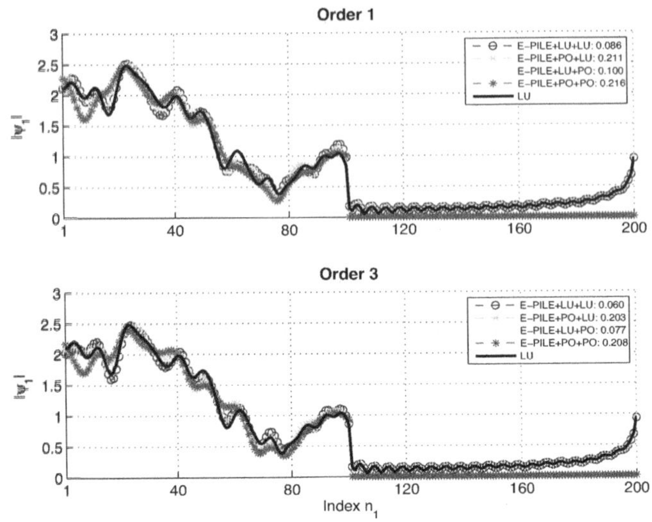

Figure 3.17. *Modulus of the currents on scatterer 1 (bottom plate) versus the index n_1. The samples $n_1 \in [1; 100]$ correspond to the upper side of the plate whereas the samples $n_1 \in [101; 200]$ correspond to the lower side*

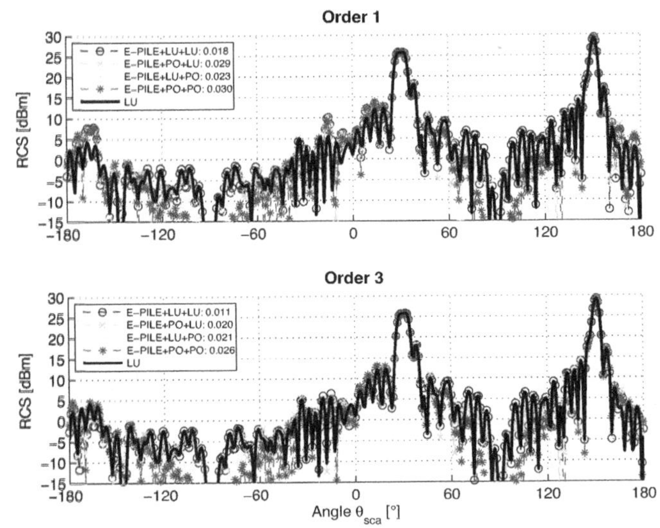

Figure 3.18. *RCS versus the scattering angle θ_{sca}. The parameters are the same as in Figure 3.17*

– “E-PILE+LU+LU” means that currents are computed from E-PILE, in which the local interactions on scatterers 1 and 2 are both computed from an LU inversion.

– “E-PILE+PO+LU” means that currents are computed from E-PILE, in which the local interactions on scatterers 1 and 2 are computed from PO and an LU inversion, respectively.

– “E-PILE+LU+PO” means that currents are computed from E-PILE, in which the local interactions on scatterers 1 and 2 are computed from an LU inversion and PO, respectively.

– “E-PILE+PO+PO” means that currents are computed from E-PILE, in which the local interactions on scatterers 1 and 2 are both computed from PO.

– “LU” means that the currents are computed from a direct LU inversion.

In addition, the value given in the figure is the corresponding residual relative error between “LU” and the corresponding method.

As the order $P_{\text{E-PILE}}$ (given in the title of each subfigure) increases, Figures 3.17 and 3.18 show that the results computed from E-PILE converge toward those obtained from LU and that the currents evaluated from “E-PILE+PO+PO” weakly change in the illuminated zone, whereas they vanish in the shadow zone and do not vary with $P_{\text{E-PILE}}$. E-PILE+PO+PO gives satisfactory results in comparison to E-PILE and when an LU inversion is applied to calculate the local interactions on either the scatterer 1 or scatterer 2, the results are similar to those obtained from “E-PILE+PO+PO”.

We consider the scenario depicted in Figure 3.16b): two PC circular cylinders of radius $a_1 = a_2 = 2.5\lambda_0$ and centers $C_{1,2} = (\mp 5\lambda_0, 0)$. The incident angle is $\theta_{\text{inc}} = 0$, the polarization is TM and $N_{\lambda_0} = 10$.

Figures 3.19 and 3.20 show the same results as in Figures 3.17 and 3.18, respectively, but they consider the scene depicted in Figure 3.16b). As we can see in Figure 3.19, for $\theta_1 \in [\pi/2; 3\pi/2]$, the field evaluated from PO is constant (2 for $\theta_1 \in [\pi/2; \pi]$ and 0 for $\theta_1 \in [\pi; 3\pi/2]$) because a point on scatterer 1 cannot interact with a point of scatterer 2, due to the shadow that is taken into account via the coupling matrices. In this region, E-PILE+LU+LU predicts a non-zero current due to the creeping waves. Figure 3.20 also shows that E-PILE+PO+PO gives satisfactory results and that the residual errors of

E-PILE+PO+LU and E-PILE+LU+PO are the same, which is consistent with the reciprocity principle since cylinder 1 is symmetrical to cylinder 2 with respect to the straight line $x = 0$, as $\theta_{\text{inc}} = 0$.

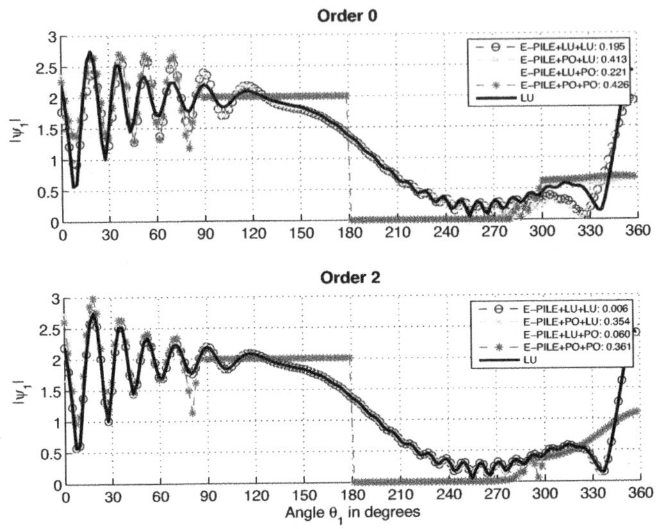

Figure 3.19. *Modulus of the currents on scatterer 1 (left cylinder) versus the polar angle* θ_1. $\theta_1 = 0$ *corresponds to the point of coordinates* $(0, -2.5\lambda_0)$ *as in Figure 3.16b*

The last scenario is depicted in Figure 3.16c): a PC plate above a rough surface with a plate: $L_1 = 10\lambda_0$, $\alpha_1 = 0$ and $C_1 = \{0, 5\}\lambda_0$, and a rough surface: $L_2 = 300\lambda_0$, $\sigma_z = 0.5\lambda_0$, $L_c = 5\lambda_0$, the surface height autocorrelation function is Gaussian and the parameter of the Thorsos wave (defined from equation [1.32]) is $g = L_1/6$. The incident angle is $\theta_{\text{inc}} = 30^\text{o}$, the polarization is TE and $N_{\lambda_0} = 10$. Figure 3.21 shows the corresponding NRCS in dB scale. As we can see, E-PILE+PO+PO and E-PILE+PO+LU predict good results in comparison to E-PILE+LU+LU and when PO is applied onto the rough surface (E-PILE+LU+PO), the results are not as good. In general [KUB 08, KUB 10c, KUB 10b], for a PC object (cylinder, plate and cross) above a rough surface, E-PILE+PO+LU gave satisfactory results. In addition, for the calculation of the local interactions on the rough surface, FB can be applied instead of a direct LU inversion, reducing the complexity without loss in precision.

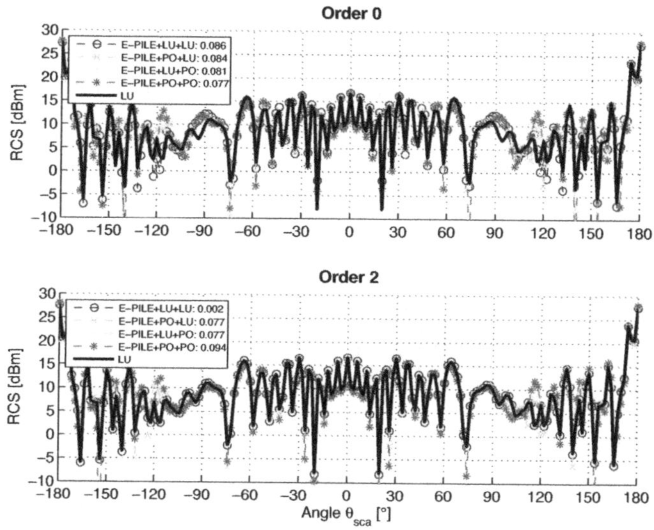

Figure 3.20. *RCS versus the scattering angle θ_{sca}. The parameters are the same as in Figure 3.19*

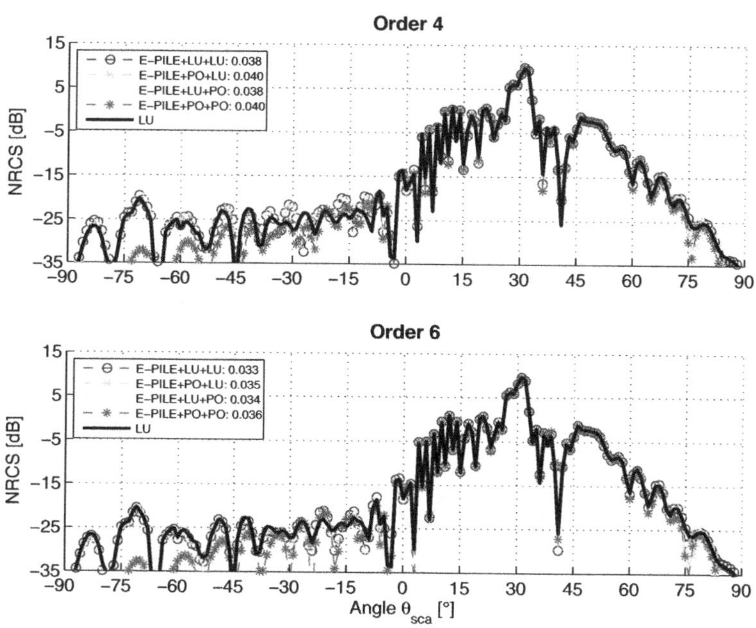

Figure 3.21. *NRCS in dB versus the scattering angle θ_{sca}. The parameters are the same as in Figure 3.16c*

Figures 3.22 and 3.23 show the same results as in Figures 3.18–3.20 but for the TE polarization. As we can see, E-PILE+PO+PO gives satisfactory results in comparison to E-PILE+LU+LU and, the relative residual error indicated in the figure is larger than that obtained for the TM polarization.

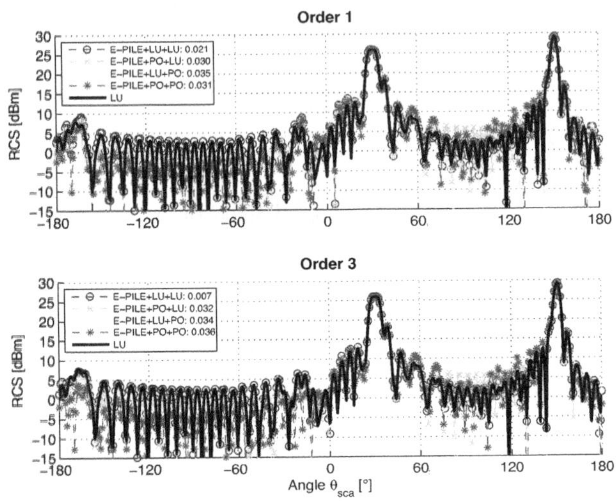

Figure 3.22. *RCS in dBm versus the scattering angle θ_{sca}. The parameters are the same as in Figure 3.18 but for the TE polarization (scenario depicted in Figure 3.16a)*

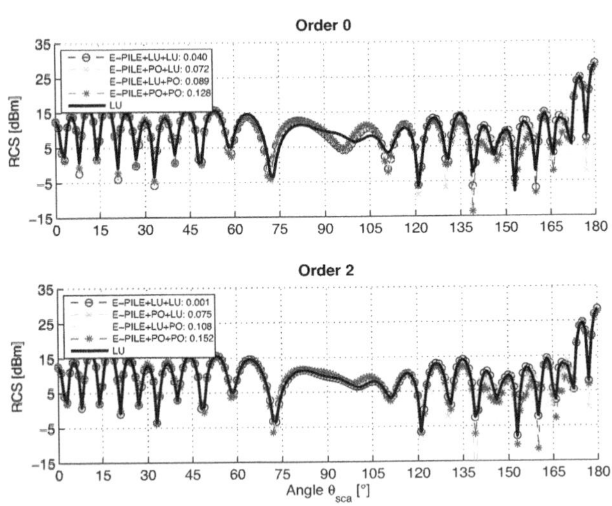

Figure 3.23. *RCS in dBm versus the scattering angle θ_{sca}. Same parameters as in Figure 3.20 but for the TE polarization (scenario depicted in Figure 3.16b)*

The E-PILE+PO+PO method can be extended to two dielectric scatterers, but the complexity of programming significantly increases.

3.5. Conclusion

In this chapter, an efficient numerical method has been presented, from the integral equations, to calculate the field scattered by two illuminated scatterers. The E-PILE method starts from the MoM and the impedance matrix of the two scatterers is then inverted by blocks from a Taylor series expansion of the inverse of the Schur complement. Its main interest is that it is rigorous, with a simple formulation and with a straightforward physical interpretation. Actually, this last property relies on the fact that each block of the impedance matrix is linked to a particular and quasi-independent physical process occurring during the multiple-scattering process between the two scatterers: local interactions on each interface, corresponding to the inversion of the impedance matrix of each scatterer in free space, and both upward and downward coupling. Furthermore, the E-PILE method allows us to use any fast method developed for a single interface. Here, PO and FB have been hybridized with E-PILE to decrease the complexity of E-PILE, and according to the scenario, this hybridization gives satisfactory results.

4

Scattering from Two Scatterers Where Only One Is Illuminated

4.1. Introduction

In this chapter, the field scattered by two objects is calculated, but unlike in Chapter 3, only one scatterer is illuminated. For instance, this general issue concerns the scattering from a coated object, from a stack of two rough interfaces of infinite length separating homogeneous media (a rough layer) or from an object below a rough surface of infinite length (see Figure 4.1). Numerically, it is not possible to generate a surface of infinite length. From the point of view of numerical methods, "infinite length" means that the surface is large compared to the electromagnetic wavelength, and for random rough surfaces, large compared to the correlation length. In addition, the incident field and the surface currents on the surface edges must both vanish.

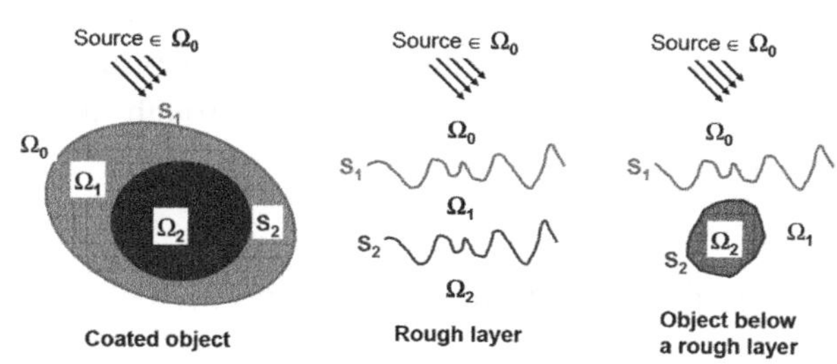

Figure 4.1. *Scattering from two scatterers where only one is illuminated. The incident wave is defined inside the medium* Ω_0

Following the same idea as in Chapter 3, the efficient propagation-inside-layer expansion (PILE) method, developed by Déchamps *et al.* [DÉC 06], is presented and applied for different scenarios. This method, originally devoted to the scattering from a random rough layer, has the main advantage that the resolution of the linear system (obtained from the method of moments (MoM)) is divided into different steps. Two steps are dedicated for solving the local interactions (computed from inverting a matrix), which can be done using efficient methods valid for a single rough interface, such as forward–backward spectral acceleration FB-SA and banded matrix iterative method/canonical grid (BMIA/CAG). Two steps are dedicated to solving the coupling interactions (com puted from a matrix-vector product), which can be done by updating the previous efficient methods. This has been investigated by BMIA/CAG [DÉC 07a] and FB-SA [DÉC 07b] for a rough layer and by Bourlier *et al.* [BOU 08] for an object located below a rough surface, in which the FB-SA is applied to calculate the local interactions on the rough surface. In addition, the order of the PILE method corresponds to the number of reflections between the two scatterers.

The applications of the general issue are numerous and it is not possible to present an exhaustive review. For a partial review, see [JOH 02, WAN 03, DÉC 06, DÉC 07a, DÉC 07b, BOU 08, AHM 08, PAW 11, FIA 12, ZOU 11, ZOU 13] (references therein).

The purpose of this chapter is to present the efficient PILE method and to explain how the calculation of the local interactions on each scatterer can be accelerated from the physical optics (PO) and FB methods, valid for a single scatterer. Section 4.2 presents the integral equations and the MoM. In section 4.3, the PILE algorithm is presented and section 4.4 presents how PO and FB can be combined with PILE. Several numerical examples are discussed, such as the scattering from a coated cylinder, a random rough layer or an object below a random rough surface.

4.2. Integral equations and method of moments

4.2.1. *Integral equations*

The problem to be solved is shown in Figure 4.1. A source illuminating the upper scatterer (object or rough surface) is defined inside the medium Ω_0 and the second scatterer is embedded inside the medium Ω_1 separating media Ω_0 and Ω_2. The medium Ω_2 is bounded by the surface S_2 of scatterer 2.

From the Helmholtz wave propagation equation [1.13] in each medium, from the definition of the Green function [1.14] and from the far radiation condition, the use of the Green theorem leads to [DÉC 06]:

$$\begin{cases} \psi[\boldsymbol{r}'] & \text{if } \boldsymbol{r}' \in \Omega_0 \\ 0 & \text{otherwise} \end{cases} = \psi_{\text{inc}}(\boldsymbol{r}') + \int_{S_1} \left[\psi(\boldsymbol{r}) \frac{\partial g_0(\boldsymbol{r}, \boldsymbol{r}')}{\partial n} - g_0(\boldsymbol{r}, \boldsymbol{r}') \frac{\partial \psi(\boldsymbol{r})}{\partial n} \right] dS, \tag{4.1}$$

$$\begin{aligned} \begin{cases} \psi[\boldsymbol{r}'] & \text{if } \boldsymbol{r}' \in \Omega_1 \\ 0 & \text{otherwise} \end{cases} = & -\int_{S_1} \left[\psi(\boldsymbol{r}) \frac{\partial g_1(\boldsymbol{r}, \boldsymbol{r}')}{\partial n} - g_1(\boldsymbol{r}, \boldsymbol{r}') \frac{\partial \psi(\boldsymbol{r})}{\partial n} \right] dS \\ & + \int_{S_2} \left[\psi(\boldsymbol{r}) \frac{\partial g_1(\boldsymbol{r}, \boldsymbol{r}')}{\partial n} - g_1(\boldsymbol{r}, \boldsymbol{r}') \frac{\partial \psi(\boldsymbol{r})}{\partial n} \right] dS, \end{aligned} \tag{4.2}$$

$$\begin{cases} \psi(\boldsymbol{r}') & \text{if } \boldsymbol{r}' \in \Omega_2 \\ 0 & \text{otherwise} \end{cases} = -\int_{S_2} \left[\psi(\boldsymbol{r}) \frac{\partial g_2(\boldsymbol{r}, \boldsymbol{r}')}{\partial n} - g_2(\boldsymbol{r}, \boldsymbol{r}') \frac{\partial \psi(\boldsymbol{r})}{\partial n} \right] dS, \tag{4.3}$$

in which g_i is the Green function inside the medium Ω_i ($i = \{1, 2, 3\}$).

To obtain the integral equations, $\boldsymbol{r}' \to S_1^+$ (tends to S_1 from above) in equations [4.1] (case $\boldsymbol{r}' \in \Omega_0$) and [4.2] (case $\boldsymbol{r}' \notin \Omega_1$) and $\boldsymbol{r}' \to S_2^+$ (tends to S_2 from above) in equations [4.3] (case $\boldsymbol{r}' \notin \Omega_2$) and [4.2] (case $\boldsymbol{r}' \in \Omega_1$). Then, from the boundary conditions applied onto scatterers 1 and 2 expressed as:

$$\begin{cases} \psi(\boldsymbol{r})|_{\boldsymbol{r}\in\Omega_0} = \psi(\boldsymbol{r})|_{\boldsymbol{r}\in\Omega_1} & \forall \boldsymbol{r} \in S_1 \\ \left.\dfrac{\partial \psi(\boldsymbol{r})}{\partial n}\right|_{\boldsymbol{r}\in\Omega_0} = \rho_{01} \left.\dfrac{\partial \psi(\boldsymbol{r})}{\partial n}\right|_{\boldsymbol{r}\in\Omega_1} & \forall \boldsymbol{r} \in S_1 \\ \psi(\boldsymbol{r})|_{\boldsymbol{r}\in\Omega_1} = \psi(\boldsymbol{r})|_{\boldsymbol{r}\in\Omega_2} & \forall \boldsymbol{r} \in S_2 \\ \left.\dfrac{\partial \psi(\boldsymbol{r})}{\partial n}\right|_{\boldsymbol{r}\in\Omega_1} = \rho_{12} \left.\dfrac{\partial \psi(\boldsymbol{r})}{\partial n}\right|_{\boldsymbol{r}\in\Omega_2} & \forall \boldsymbol{r} \in S_2 \end{cases}, \tag{4.4}$$

we obtain the following four integral equations:

$$\psi_{\text{inc}}(\boldsymbol{r}') = \underbrace{+\frac{1}{2}\psi_1(\boldsymbol{r}') - \fint_{S_1} \frac{\partial g_0(\boldsymbol{r},\boldsymbol{r}')}{\partial n}\psi_1(\boldsymbol{r})dS}_{\Rightarrow \bar{\boldsymbol{A}}_1} + \underbrace{\int_{S_1} g_0(\boldsymbol{r},\boldsymbol{r}')\frac{\partial \psi_1(\boldsymbol{r})}{\partial n}dS}_{\Rightarrow \bar{\boldsymbol{B}}_1}$$

$$\forall(\boldsymbol{r},\boldsymbol{r}') \in S_1, \tag{4.5}$$

$$0 = \underbrace{-\frac{1}{2}\psi_1(\boldsymbol{r}') - \fint_{S_1} \frac{\partial g_1(\boldsymbol{r},\boldsymbol{r}')}{\partial n}\psi_1(\boldsymbol{r})dS}_{\Rightarrow \bar{\boldsymbol{C}}_1} + \frac{1}{\rho_{01}}\underbrace{\int_{S_1} g_1(\boldsymbol{r},\boldsymbol{r}')\frac{\partial \psi_1(\boldsymbol{r})}{\partial n}dS}_{\Rightarrow \bar{\boldsymbol{D}}_1}$$

$$+ \underbrace{\int_{S_2} \frac{\partial g_1(\boldsymbol{r},\boldsymbol{r}')}{\partial n}\psi_2(\boldsymbol{r})dS}_{\Rightarrow \bar{\boldsymbol{A}}_{21}} - \underbrace{\int_{S_2} g_1(\boldsymbol{r},\boldsymbol{r}')\frac{\partial \psi_2(\boldsymbol{r})}{\partial n}dS}_{\Rightarrow \bar{\boldsymbol{B}}_{21}}$$

$$\forall\, \boldsymbol{r}' \in S_1, \tag{4.6}$$

$$0 = \underbrace{+\frac{1}{2}\psi_2(\boldsymbol{r}') - \fint_{S_2} \frac{\partial g_1(\boldsymbol{r},\boldsymbol{r}')}{\partial n}\psi_2(\boldsymbol{r})dS}_{\Rightarrow \bar{\boldsymbol{A}}_2} + \underbrace{\int_{S_2} g_1(\boldsymbol{r},\boldsymbol{r}')\frac{\partial \psi_2(\boldsymbol{r})}{\partial n}dS}_{\Rightarrow \bar{\boldsymbol{B}}_2}$$

$$+ \underbrace{\int_{S_1} \frac{\partial g_1(\boldsymbol{r},\boldsymbol{r}')}{\partial n}\psi_1(\boldsymbol{r})dS}_{\Rightarrow \bar{\boldsymbol{A}}_{12}} - \underbrace{\int_{S_1} g_1(\boldsymbol{r},\boldsymbol{r}')\frac{\partial \psi_1(\boldsymbol{r})}{\partial n}dS}_{\Rightarrow \bar{\boldsymbol{B}}_{12}}\frac{1}{\rho_{01}}$$

$$\forall\, \boldsymbol{r}' \in S_2, \tag{4.7}$$

$$0 = \underbrace{-\frac{1}{2}\psi_2(\boldsymbol{r}') - \fint_{S_2} \frac{\partial g_2(\boldsymbol{r},\boldsymbol{r}')}{\partial n}\psi_2(\boldsymbol{r})dS}_{\Rightarrow \bar{\boldsymbol{C}}_2} + \frac{1}{\rho_{12}}\underbrace{\int_{S_2} g_2(\boldsymbol{r},\boldsymbol{r}')\frac{\partial \psi_2(\boldsymbol{r})}{\partial n}dS}_{\Rightarrow \bar{\boldsymbol{D}}_2}$$

$$\forall(\boldsymbol{r},\boldsymbol{r}') \in S_2, \tag{4.8}$$

where $\{\psi_j, \partial\psi_j/\partial n\}$ are the currents on the surface S_j ($j = \{1,2\}$). In addition, $\rho_{ij} = 1$ ($i = \{0,1\}$ and $j \neq i$) in the TE polarization and $\rho_{ij} = \epsilon_i/\epsilon_j$ in the TM polarization, where ϵ_i is the permittivity of medium Ω_i. The symbol $\fint$ stands for the Cauchy principal value, which means that the case $\boldsymbol{r} = \boldsymbol{r}'$ is not accounted for in the calculation of the integral. In addition, letting $\int_{S_j} dS = \fint dS + \int_{P_j} dS$ (where P_j is a piece), it is important to note that for $\boldsymbol{r} = \boldsymbol{r}' \in S_j^+$ (at the upper limit of the surface), $\int_{P_j} \psi_j(\boldsymbol{r}) \frac{\partial g_j(\boldsymbol{r},\boldsymbol{r}')}{\partial n} dS = +\psi_j(\boldsymbol{r}')/2$, and for $\boldsymbol{r} = \boldsymbol{r}' \in S_j^-$ (at the lower limit of the surface), $\int_{P_j} \psi_j(\boldsymbol{r}) \frac{\partial g_j(\boldsymbol{r},\boldsymbol{r}')}{\partial n} dS = -\psi_j(\boldsymbol{r}')/2$. The sign difference is due to the sense of the normal to the surface S_j.

From equations [4.5]–[4.8], the currents $\{\psi_j, \partial\psi_j/\partial n\}$ on the surface S_j can be computed from the MoM.

4.2.2. *Method of moments*

In this section, the MoM is applied with the point-matching method to convert the integral equations [4.5]–[4.8] into a linear system. This method was thoroughly explained in section 1.3.

Equations [4.5]–[4.8] lead to a coupled system. From the MoM, the equations are discretized on each surface of the scatterer, leading to the linear system $\bar{\boldsymbol{Z}}\boldsymbol{X} = \boldsymbol{b}$. The unknown vector is then:

$$\boldsymbol{X} = \begin{bmatrix} \boldsymbol{X}_1 \\ \boldsymbol{X}_2 \end{bmatrix}, \tag{4.9}$$

where the components of the vectors $\boldsymbol{X}_1$ and $\boldsymbol{X}_2$ are the currents discretized on the surfaces S_1 and S_2, respectively. They are written as:

$$\boldsymbol{X}_1 = \left[\psi_1(\boldsymbol{r}_1) \ldots \psi_1(\boldsymbol{r}_{N_1}) \ \frac{\partial\psi_1(\boldsymbol{r}_1)}{\partial n} \ldots \frac{\partial\psi_1(\boldsymbol{r}_{N_1})}{\partial n} \right]^{\mathrm{T}} \quad \boldsymbol{r}_{p\in[1;N_1]} \in S_1, \tag{4.10}$$

$$\boldsymbol{X}_2 = \left[\psi_2(\boldsymbol{r}_1) \ldots \psi_2(\boldsymbol{r}_{N_2}) \ \frac{\partial\psi_1(\boldsymbol{r}_2)}{\partial n} \ldots \frac{\partial\psi_2(\boldsymbol{r}_{N_2})}{\partial n} \right]^{\mathrm{T}} \quad \boldsymbol{r}_{p\in[1;N_2]} \in S_2, \tag{4.11}$$

where the symbol T stands for the transpose operator and N_i is the number of samples on the surface S_i. Then, the length of the vector $\boldsymbol{X}_i$ is $2N_i$.

The vector $\boldsymbol{b}$ of length $2(N_1+N_2)$ is the incident field discretized on the surfaces S_1 and S_2. It is defined as:

$$\boldsymbol{b}=\begin{bmatrix}\boldsymbol{b}_1\\ \boldsymbol{b}_2\end{bmatrix}=\begin{bmatrix}\underbrace{\psi_{\text{inc}}(\boldsymbol{r}_1)\ldots\psi_{\text{inc}}(\boldsymbol{r}_{N_1})\ \underbrace{0\ldots0}_{N_1\text{ times}}}_{\boldsymbol{b}_1^{\mathrm{T}},\boldsymbol{r}\in S_1}\ \underbrace{\underbrace{0\ldots0}_{2N_2\text{ times}}}_{\boldsymbol{b}_2^{\mathrm{T}},\boldsymbol{r}\in S_2}\end{bmatrix}^{\mathrm{T}}. \tag{4.12}$$

The impedance matrix $\bar{\boldsymbol{Z}}$ of size $2(N_1+N_2)\times 2(N_1+N_2)$ is expressed as:

$$\bar{\boldsymbol{Z}}=\begin{bmatrix}\bar{\boldsymbol{A}}_1 & \bar{\boldsymbol{B}}_1 & \bar{0} & \bar{0}\\ \bar{\boldsymbol{C}}_1 & \dfrac{1}{\rho_{01}}\bar{\boldsymbol{D}}_1 & \bar{\boldsymbol{A}}_{21} & \bar{\boldsymbol{B}}_{21}\\ \bar{\boldsymbol{A}}_{12} & \dfrac{1}{\rho_{01}}\bar{\boldsymbol{B}}_{12} & \bar{\boldsymbol{A}}_2 & \bar{\boldsymbol{B}}_2\\ \bar{0} & \bar{0} & \bar{\boldsymbol{C}}_2 & \dfrac{1}{\rho_{12}}\bar{\boldsymbol{D}}_2\end{bmatrix}=\begin{bmatrix}\bar{\boldsymbol{Z}}_1 & \bar{\boldsymbol{Z}}_{21}\\ \bar{\boldsymbol{Z}}_{12} & \bar{\boldsymbol{Z}}_2\end{bmatrix}, \tag{4.13}$$

where

$$\bar{\boldsymbol{Z}}_1=\begin{bmatrix}\bar{\boldsymbol{A}}_1 & \bar{\boldsymbol{B}}_1\\ \bar{\boldsymbol{C}}_1 & \dfrac{1}{\rho_{01}}\bar{\boldsymbol{D}}_1\end{bmatrix},\quad \bar{\boldsymbol{Z}}_2=\begin{bmatrix}\bar{\boldsymbol{A}}_2 & \bar{\boldsymbol{B}}_2\\ \bar{\boldsymbol{C}}_2 & \dfrac{1}{\rho_{12}}\bar{\boldsymbol{D}}_2\end{bmatrix}, \tag{4.14}$$

and

$$\bar{\boldsymbol{Z}}_{21}=\begin{bmatrix}\bar{0} & \bar{0}\\ \bar{\boldsymbol{A}}_{21} & \bar{\boldsymbol{B}}_{21}\end{bmatrix},\quad \bar{\boldsymbol{Z}}_{12}=\begin{bmatrix}\bar{\boldsymbol{A}}_{12} & \dfrac{1}{\rho_{01}}\bar{\boldsymbol{B}}_{12}\\ \bar{0} & \bar{0}\end{bmatrix}. \tag{4.15}$$

The impedance matrix $\bar{\boldsymbol{Z}}_i$ of size $2N_i\times 2N_i$ is the impedance matrix of scatterer i in free space, where N_i is the number of the samples on scatterer i.

In comparison to Chapter 3, in which the two scatterers are illuminated ($\boldsymbol{b}_1$ and $\boldsymbol{b}_2$ different of $\mathbf{0}$), for scatterer 2, ρ_{02} is substituted for ρ_{12} (equations [3.13] and [4.14]) because scatterer 2 is embedded in medium Ω_1. In addition, the coupling matrices $\bar{\boldsymbol{Z}}_{21}$ and $\bar{\boldsymbol{Z}}_{12}$ between scatterers 1 and 2 slightly differ, because the wave number is k_1 instead of k_0 as mentioned in Chapter 3.

From equations [3.15] and [3.16], the elements of the matrices $\{\bar{\boldsymbol{A}}_1, \bar{\boldsymbol{B}}_1, \bar{\boldsymbol{C}}_1, \bar{\boldsymbol{D}}_1\}$ are:

$$A_{1,mn} = \begin{cases} -\dfrac{jk_0 v_n|\Delta_n|}{4} \dfrac{\mathrm{H}_1^{(1)}\left(k_0\|\boldsymbol{r}_n - \boldsymbol{r}_m\|\right)}{\|\boldsymbol{r}_n - \boldsymbol{r}_m\|} \\ \qquad \times \left[\gamma_n(x_n - x_m) - (z_n - z_m)\right] \text{ for } m \neq n \;, \\ +\dfrac{1}{2} - \dfrac{v_n|\Delta_n|}{4\pi} \dfrac{\gamma'(x_n)}{1+\gamma^2(x_n)} \qquad \text{for } m = n \end{cases} \quad [4.16]$$

$$B_{1,mn} = \frac{j|\Delta_n|\sqrt{1+\gamma_n^2}}{4} \begin{cases} \mathrm{H}_0^{(1)}\left(k_0\|\boldsymbol{r}_n - \boldsymbol{r}_m\|\right) & \text{for } m \neq n \\ 1 + \dfrac{2j}{\pi} \ln\left(0.164 k_0 \sqrt{1+\gamma_n^2}|\Delta_n|\right) & \text{for } m = n \end{cases}, \quad [4.17]$$

$$C_{1,mn} = \begin{cases} -\dfrac{jk_1 v_n|\Delta_n|}{4} \dfrac{\mathrm{H}_1^{(1)}\left(k_1\|\boldsymbol{r}_n - \boldsymbol{r}_m\|\right)}{\|\boldsymbol{r}_n - \boldsymbol{r}_m\|} \\ \qquad \times \left[\gamma_n(x_n - x_m) - (z_n - z_m)\right] \text{ for } m \neq n \;, \\ -\dfrac{1}{2} - \dfrac{v_n|\Delta_n|}{4\pi} \dfrac{\gamma'(x_n)}{1+\gamma^2(x_n)} \qquad \text{for } m = n \end{cases} \quad [4.18]$$

$$D_{1,mn} = \frac{j|\Delta_n|\sqrt{1+\gamma_n^2}}{4} \begin{cases} \mathrm{H}_0^{(1)}\left(k_1\|\boldsymbol{r}_n - \boldsymbol{r}_m\|\right) & \text{for } m \neq n \\ 1 + \dfrac{2j}{\pi} \ln\left(0.164 k_1 \sqrt{1+\gamma_n^2}|\Delta_n|\right) & \text{for } m = n \end{cases}, \quad [4.19]$$

where $\boldsymbol{r}_n = (x_n, z_n) \in S_1$, $\boldsymbol{r}_m = (x_m, z_m) \in S_1$, $\gamma_n = dz_n/dx_n$, Δ_n the sampling step, $v_n = \mathrm{sgn}(\hat{\boldsymbol{n}}_n \cdot \hat{\boldsymbol{z}})$ ($\hat{\boldsymbol{n}}_n$ is the vector normal to the surface at the point $\boldsymbol{r}_n$), $\mathrm{H}_0^{(1)}$ the zeroth-order Hankel function of the first kind and $\mathrm{H}_1^{(1)}$ its derivative. The elements of the matrices $\{\bar{\boldsymbol{A}}_2, \bar{\boldsymbol{B}}_2, \bar{\boldsymbol{C}}_2, \bar{\boldsymbol{D}}_2\}$ are obtained from those of $\{\bar{\boldsymbol{A}}_1, \bar{\boldsymbol{B}}_1, \bar{\boldsymbol{C}}_1, \bar{\boldsymbol{D}}_1\}$ by substituting the wave numbers $\{k_0, k_0, k_1, k_1\}$ for $\{k_1, k_1, k_2, k_2\}$, respectively.

From equation [3.17], the elements of the coupling matrices $\{\bar{\boldsymbol{A}}_{12}, \bar{\boldsymbol{B}}_{12}, \bar{\boldsymbol{A}}_{21}, \bar{\boldsymbol{B}}_{21}\}$ are:

$$\left\{\begin{array}{l} A_{12,mn} = -\dfrac{jk_1 v_{1,n}|\Delta_{1,n}|}{4} \dfrac{\mathrm{H}_1^{(1)}\left(k_1\|\boldsymbol{r}_{1,n}-\boldsymbol{r}_{2,m}\|\right)}{\|\boldsymbol{r}_{1,n}-\boldsymbol{r}_{2,m}\|} \\ \qquad \times\left[\gamma_{1,n}(x_{1,n}-x_{2,m})-(z_{1,n}-z_{2,m})\right] \\ \\ A_{21,mn} = -\dfrac{jk_1 v_{2,n}|\Delta_{2,n}|}{4} \dfrac{\mathrm{H}_1^{(1)}\left(k_1\|\boldsymbol{r}_{2,n}-\boldsymbol{r}_{1,m}\|\right)}{\|\boldsymbol{r}_{2,n}-\boldsymbol{r}_{1,m}\|} \\ \qquad \times\left[\gamma_{2,n}(x_{2,n}-x_{1,m})-(z_{2,n}-z_{1,m})\right] \\ \\ B_{12,mn} = \frac{j|\Delta_{1,n}|\sqrt{1+\gamma_{1,n}^2}}{4}\mathrm{H}_0^{(1)}\left(k_1\|\boldsymbol{r}_{1,n}-\boldsymbol{r}_{2,m}\|\right) \\ \\ B_{21,mn} = \frac{j|\Delta_{2,n}|\sqrt{1+\gamma_{2,n}^2}}{4}\mathrm{H}_0^{(1)}\left(k_1\|\boldsymbol{r}_{2,n}-\boldsymbol{r}_{1,m}\|\right) \end{array}\right., \qquad [4.20]$$

for all (n, m). The subscripts "$1, n$" and "$2, m$" refer to a point on scatterers 1 and 2, respectively.

From the knowledge of the currents $\{\psi_i, \partial\psi_i/\partial n\}$ on scatterers S_i ($i = \{1, 2\}$), the scattered field $\psi_{\text{sca},\nu}$ inside the medium Ω_ν ($\nu = \{0, 1, 2\}$) is computed from the Huygens' principle as:

$$\left\{\begin{array}{l} \psi_{\text{sca},0}(\boldsymbol{r}') = \displaystyle\int_{S_1}\left[\psi_0(\boldsymbol{r})\frac{\partial g_0(\boldsymbol{r},\boldsymbol{r}')}{\partial n} - g_0(\boldsymbol{r},\boldsymbol{r}')\frac{\partial\psi_0(\boldsymbol{r})}{\partial n}\right]dS \\ \psi_{\text{sca},1}(\boldsymbol{r}') = \displaystyle\sum_{p=1}^{p=2} w_p\int_{S_p}\left[\psi_p(\boldsymbol{r})\frac{\partial g_1(\boldsymbol{r},\boldsymbol{r}')}{\partial n} - g_1(\boldsymbol{r},\boldsymbol{r}')\frac{\partial\psi_p(\boldsymbol{r})}{\partial n}\right]dS, \\ \psi_{\text{sca},2}(\boldsymbol{r}') = -\displaystyle\int_{S_2}\left[\psi_2(\boldsymbol{r})\frac{\partial g_2(\boldsymbol{r},\boldsymbol{r}')}{\partial n} - g_2(\boldsymbol{r},\boldsymbol{r}')\frac{\partial\psi_2(\boldsymbol{r})}{\partial n}\right]dS \end{array}\right. \qquad [4.21]$$

with $w_1 = -1$, $w_2 = +1$ and $g_\nu(\boldsymbol{r},\boldsymbol{r}') = (j/4)\mathrm{H}_0^{(1)}(k_\nu\|\boldsymbol{r}'-\boldsymbol{r}\|)$.

4.2.3. *Case for which scatterer 2 is perfectly conducting*

If scatterer 2 is assumed to be perfectly conducting, the impedance matrix $\bar{\boldsymbol{Z}}_2$ can be simplified. For the TE polarization (Dirichlet boundary condition),

ψ_2 vanishes on the surface and the only unknown on the surface is $\partial\psi_2/\partial n$. Then

$$\text{TE: } \bar{\boldsymbol{Z}}_2 = \bar{\boldsymbol{B}}_2,\ \boldsymbol{X}_2 = \frac{\partial\psi_2}{\partial n}. \qquad [4.22]$$

For the TM polarization (Neumann boundary condition), $\partial\psi_2/\partial n$ vanishes on the surface and the only unknown on the surface is ψ_2. Then

$$\text{TM: } \bar{\boldsymbol{Z}}_2 = \bar{\boldsymbol{A}}_2,\ \boldsymbol{X}_2 = \psi_2. \qquad [4.23]$$

In addition, the coupling matrices are simplified as:

$$\begin{cases} \text{TE: } \bar{\boldsymbol{Z}}_{12} = \begin{bmatrix} \bar{\boldsymbol{A}}_{12} & \dfrac{1}{\rho_{01}}\bar{\boldsymbol{B}}_{12} \end{bmatrix} \ \bar{\boldsymbol{Z}}_{21} = \begin{bmatrix} \bar{\boldsymbol{0}} \\ \bar{\boldsymbol{B}}_{21} \end{bmatrix} \\ \text{TM: } \bar{\boldsymbol{Z}}_{12} = \begin{bmatrix} \bar{\boldsymbol{A}}_{12} & \dfrac{1}{\rho_{01}}\bar{\boldsymbol{B}}_{12} \end{bmatrix} \ \bar{\boldsymbol{Z}}_{21} = \begin{bmatrix} \bar{\boldsymbol{0}} \\ \bar{\boldsymbol{A}}_{21} \end{bmatrix} \end{cases}. \qquad [4.24]$$

4.2.4. *Numerical results*

4.2.4.1. *Case of a coated circular cylinder and comparison with the analytical solution*

For a coated circular cylinder, that is two concentric circular cylinders of radii a_1 and $a_2 < a_1$ separating homogeneous media $(\Omega_0, \Omega_1, \Omega_2)$, as discussed in section 2.3, the exact solution of the scattered field can be obtained by using the Bessel functions.

Inside medium Ω_0, from equation [2.37], the total field ψ_0 is:

$$\psi_0(r,\theta) = \sum_{n=-\infty}^{n=+\infty} \left[A_n \mathrm{J}_n(k_0 r) + B_n \mathrm{H}_n^{(1)}(k_0 r)\right] e^{-jn\theta} \text{ with } A_n = \psi_{\text{inc},0} e^{jn\theta_{\text{inc}}}. \qquad [4.25]$$

Inside medium Ω_1, from equation [2.37], the total field ψ_1 is:

$$\psi_1(r,\theta) = \sum_{n=-\infty}^{n=+\infty} \left[C_n \mathrm{J}_n(k_1 r) + D_n \mathrm{H}_n^{(1)}(k_1 r)\right] e^{-jn\theta}. \qquad [4.26]$$

Inside medium Ω_2, from equation [2.38], the total field ψ_2 is:

$$\psi_2(r,\theta) = \sum_{n=-\infty}^{n=+\infty} E_n \mathrm{J}_n(k_2 r) e^{-jn\theta}. \quad [4.27]$$

In equations [4.25]–[4.27], the four unknowns are B_n, C_n, D_n and E_n. From the boundary conditions [2.9], we have:

$$\begin{cases} \psi_0(a_1,\theta) = \psi_1(a_1,\theta) \\ \psi_1(a_2,\theta) = \psi_2(a_2,\theta) \\ \left.\dfrac{\partial\psi_0}{\partial r}\right|_{r=a_1} = \rho_{01} \left.\dfrac{\partial\psi_1}{\partial r}\right|_{r=a_1} \\ \left.\dfrac{\partial\psi_1}{\partial r}\right|_{r=a_2} = \rho_{12} \left.\dfrac{\partial\psi_2}{\partial r}\right|_{r=a_2} \end{cases}. \quad [4.28]$$

From equations [4.25]–[4.27] $\forall$ (θ, n), we have:

$$\begin{bmatrix} \mathrm{H}_n^{(1)}(k_0a_1) & -\mathrm{J}_n(k_1a_1) & -\mathrm{H}_n^{(1)}(k_1a_1) & 0 \\ k_0\mathrm{H}_n'^{(1)}(k_0a_1) & -\rho_{01}k_1\mathrm{J}_n'(k_1a_1) & -\rho_{01}k_1\mathrm{H}_n'^{(1)}(k_1a_1) & 0 \\ 0 & \mathrm{J}_n(k_1a_2) & \mathrm{H}_n^{(1)}(k_1a_2) & -\mathrm{J}_n(k_2a_2) \\ 0 & k_1\mathrm{J}_n'(k_1a_2) & k_1\mathrm{H}_n'^{(1)}(k_1a_2) & -k_2\rho_{12}\mathrm{J}_n'(k_2a_2) \end{bmatrix} \begin{bmatrix} B_n \\ C_n \\ D_n \\ E_n \end{bmatrix} = \begin{bmatrix} -A_n\mathrm{J}_n(k_0a_1) \\ -A_nk_0\mathrm{J}_n'(k_0a_1) \\ 0 \\ 0 \end{bmatrix}. \quad [4.29]$$

This linear system can be solved analytically, but for convenience it is solved numerically by inverting the matrix of size 4×4.

Figure 4.2a) shows the modulus and the phase of the currents on scatterer 1 versus the angle θ_1. Figure 4.2b) shows the modulus and the phase of the currents on scatterer 2 versus the angle θ_2. The angles θ_1 and θ_2 are defined from the horizontal direction $\hat{x}$. Figure 4.2c) shows the modulus of the total field versus the abscissa x and the height z. Figure 4.2d) shows the radar cross-section (RCS) versus the scattering angle θ_{sca}. The radii of the two concentric circular cylinders are $a_1 = 3\lambda_0$ and $a_2 = 2\lambda_0$, their centers are $C_1 = C_2 = (0,0)$, the relative permittivities of media $\{\Omega_0, \Omega_1, \Omega_2\}$ are $\{\epsilon_{r0} = 1, \epsilon_{r1} = 2, \epsilon_{r2} = 4 + 0.05j\}$, respectively. The incidence angle is $\theta_{\text{inc}} = 0$ and the polarization is TE. For the MoM, the number of samples per wavelength is

$N_{\lambda_0} = 10$ and the number of unknowns is given in parentheses in the figure. In addition, for scatterer 1 ($r = a_1$) in the figure legend, the label "Ana 1" means that the currents are computed from equation [4.25], whereas the label "Ana 2" means that the currents are computed from equation [4.26]. For scatterer 2 ($r = a_2$) in the legend, the label "Ana 1" means that the currents are computed from equation [4.26], whereas the label "Ana 2" means that the currents are computed from equation [4.27].

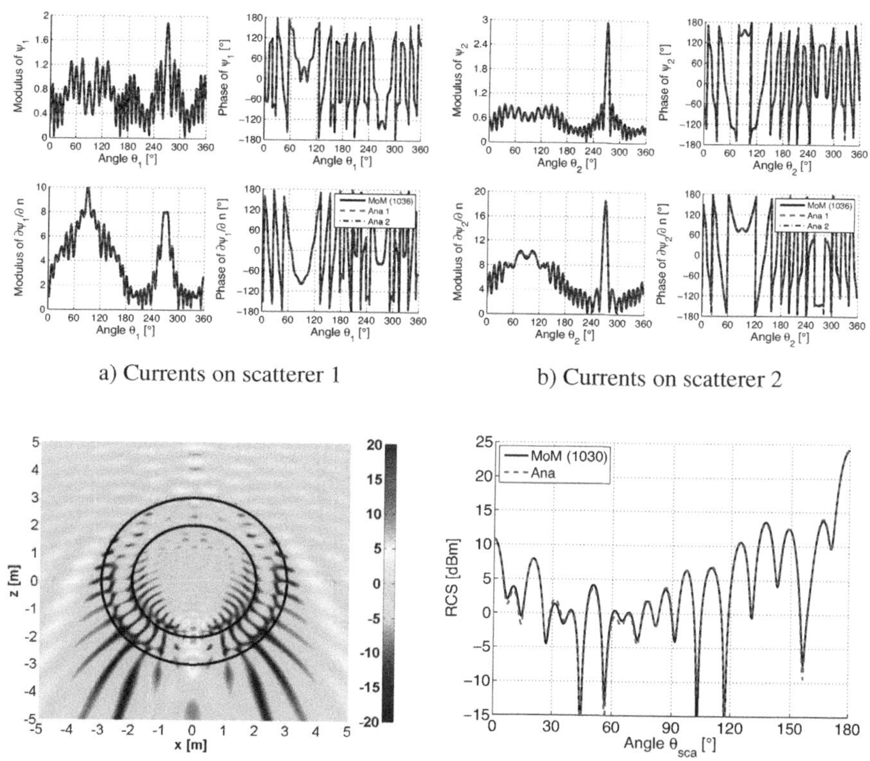

a) Currents on scatterer 1

b) Currents on scatterer 2

c) Modulus of the total field in dB scale

d) RCS

Figure 4.2. *a) Modulus and phase of the currents on scatterer 1 versus the angle θ_1. b) Modulus and phase of the currents on scatterer 2 versus the angle θ_2. c) Modulus of the total field in dB scale versus the abscissa x and the height z. d) RCS in dBm scale versus the scattering angle θ_{sca}. The radii of the two concentric circular cylinders are $a_1 = 3\lambda_0$, $a_2 = 2\lambda_0$, their centers are $C_1 = C_2 = (0, 0)$, the relative permittivities of media $\{\Omega_0, \Omega_1, \Omega_2\}$ are $\{\epsilon_{r0} = 1, \epsilon_{r1} = 2, \epsilon_{r2} = 4 + 0.05j\}$, and $\lambda_0 = 1$ m. The incidence angle is $\theta_{inc} = 0$ and the polarization is TE. For the MoM, the number of samples per wavelength is $N_{\lambda_0} = 10$*

Figure 4.3 shows the same results as in Figure 4.2 but for the TM polarization.

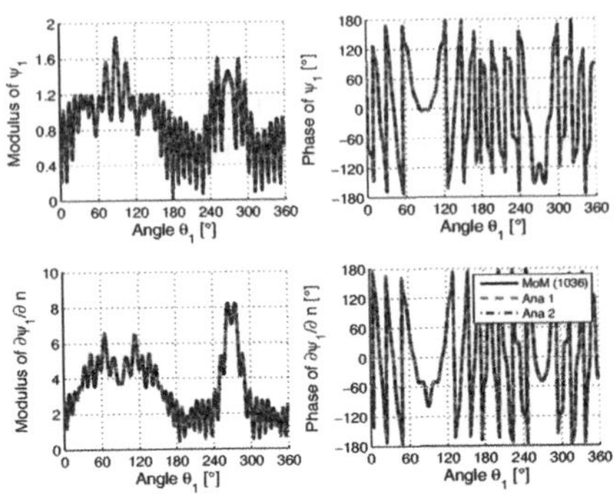

a) Currents on scatterer 1

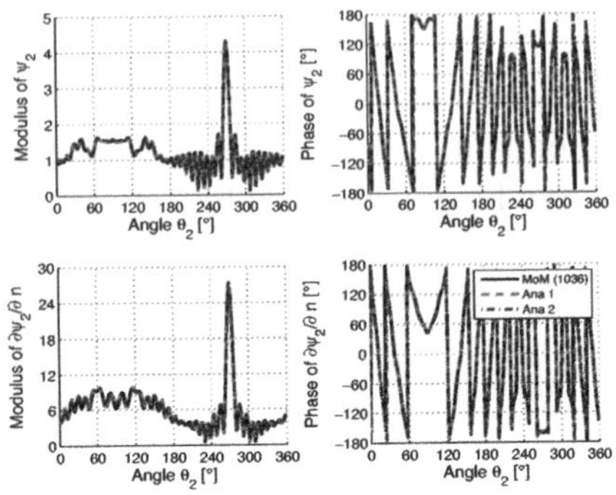

b) Currents on scatterer 2

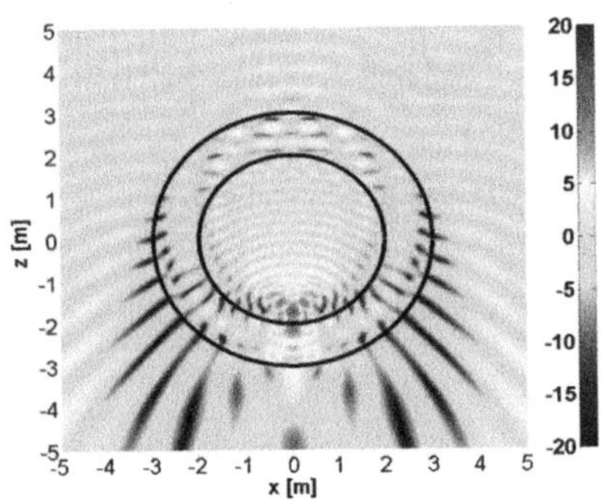

c) Modulus of the total field in dB

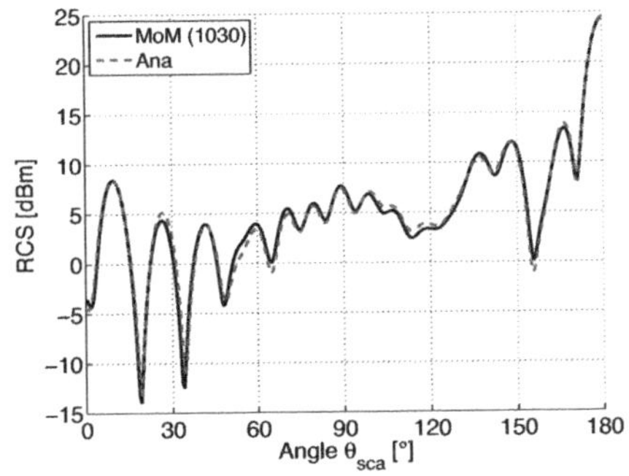

d) RCS

Figure 4.3. *Same results as in Figure 4.2 but for the TM polarization*

For the two polarizations, the results calculated from "Ana 1" and "Ana 2" perfectly match, which means that the boundary conditions are satisfied. A small difference between the MoM and the analytical solution is observed on the currents and Figures 4.2d) and 4.3d) show that this deviation has a small impact on the RCS. This deviation can be decreased by increasing N_{λ_0} (for instance, from 10 to 20).

4.2.4.2. *Case of a coated elliptical cylinder*

The MoM allows us to treat any geometry like a dielectric elliptical cylinder, which is coated eccentrically by a non-confocal dielectric elliptical cylinder. As shown in [ZOU 13], this issue can be solved by introducing the Mathieu functions and the equivalent of the Graf theorem on the Bessel functions, which significantly increases the complexity of programming. One of the advantages of the MoM is its ability to solve this issue for any incident field with only minor modifications to the program.

Figure 4.4 shows the same results as in Figures 4.2c) and 4.2d) but for two eccentric elliptical cylinders of different rotation angles $\alpha_{1,2}$ and centers $C_{1,2}$. Figure 4.5 shows the same results as in Figure 4.4 but for the TM polarization.

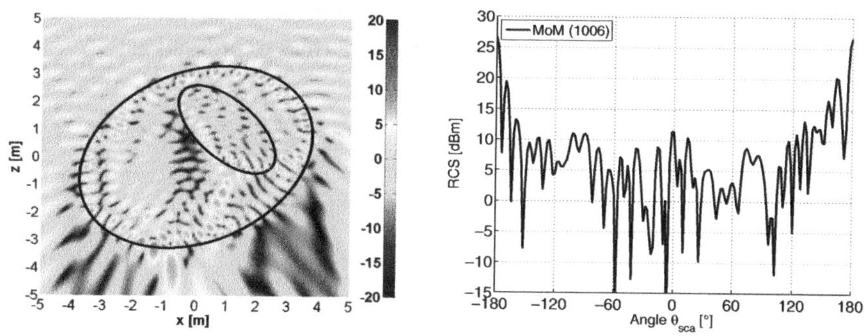

a) Modulus of the total field in dB

b) RCS

Figure 4.4. *Same results as in Figures 4.2c) and 4.2d) but* $a_1 = 4\lambda_0$, $b_1 = 3\lambda_0$, $\alpha_1 = \pi/6$, $a_2 = 2\lambda_0$, $b_2 = \lambda_0$, $\alpha_2 = -\pi/4$, $C_1 = (0, 0)$, $C_2 = (1, 1)\lambda_0$

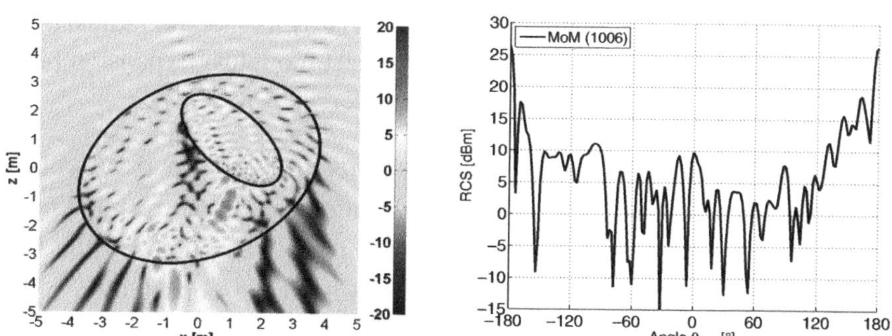

a) Modulus of the total field in dB

b) RCS

Figure 4.5. *Same results as in Figure 4.4 but for the TM polarization*

4.3. Efficient inversion of the impedance matrix: PILE method

4.3.1. *Mathematical formulation*

To efficiently solve the linear system $\bar{\boldsymbol{Z}}\boldsymbol{X} = \boldsymbol{b}$, the PILE method, developed by Déchamps *et al.* [DÉC 06], can be applied. It is based on an inversion by blocks (Schur complement [PRE 92]) of the impedance matrix [4.13]. This method was later generalized by Kubické *et al.* [KUB 08], and called extended-PILE (E-PILE), to the more general case for which the two scatterers are illuminated (see Chapter 3). Its principle is thoroughly explained in section 3.3. From equation [3.34] with $\boldsymbol{b}_2 = 0$ (scatterer 2 is not illuminated), the currents on scatterer 1, $\boldsymbol{X}_1$, are then expressed as:

$$\boldsymbol{X}_1 = \left[\sum_{p=0}^{p=P_{\text{PILE}}} \bar{\boldsymbol{M}}_c^p\right] \bar{\boldsymbol{Z}}_1^{-1}\boldsymbol{b}_1 = \sum_{p=0}^{p=P_{\text{PILE}}} \boldsymbol{Y}_1^{(p)}, \qquad [4.30]$$

in which

$$\begin{cases} \boldsymbol{Y}_1^{(0)} = \bar{\boldsymbol{Z}}_1^{-1}\boldsymbol{b}_1 & \text{for } p = 0 \\ \boldsymbol{Y}_1^{(p)} = \bar{\boldsymbol{M}}_c \boldsymbol{Y}_1^{(p-1)} & \text{for } p > 0 \end{cases}, \qquad [4.31]$$

and

$$\bar{\boldsymbol{M}}_c = \bar{\boldsymbol{Z}}_1^{-1}\bar{\boldsymbol{Z}}_{21}\bar{\boldsymbol{Z}}_2^{-1}\bar{\boldsymbol{Z}}_{12}. \qquad [4.32]$$

From equations [3.28] and [3.29], $\boldsymbol{X}_2 = \bar{\boldsymbol{V}}\boldsymbol{b}_1 = -\bar{\boldsymbol{Z}}_2^{-1}\bar{\boldsymbol{Z}}_{12}\bar{\boldsymbol{T}}\boldsymbol{b}_1 = -\bar{\boldsymbol{Z}}_2^{-1}\bar{\boldsymbol{Z}}_{12}\boldsymbol{X}_1$.

We define the norm $\|\bar{\boldsymbol{M}}_c\|$ of a complex matrix by its spectral radius, that is the modulus of its eingenvalue that has the highest modulus. Expansion [4.30] is then valid if $\|\bar{\boldsymbol{M}}_c\|$ is strictly smaller than one.

Equation [4.31] has a clear physical interpretation: the total currents on the scatterer 1 are the sum of the contributions $\boldsymbol{Y}_1^{(p)}$ corresponding to successive iterations p. In the zeroth-order term, $\bar{\boldsymbol{Z}}_1^{-1}$ accounts for the local interactions on scatterer 1, so $\boldsymbol{Y}_1^{(0)}$ corresponds to the contribution of the direct reflection on scatterer 1, without entering the medium Ω_1. In the first-order term given by $\boldsymbol{Y}_1^{(1)} = \bar{\boldsymbol{M}}_c\boldsymbol{Y}_1^{(0)}$, the matrix $\bar{\boldsymbol{Z}}_{12}$ propagates the resulting currents, $\boldsymbol{Y}_1^{(0)}$ toward scatterer 2, $\bar{\boldsymbol{Z}}_2^{-1}$ accounts for the local interactions on scatterer 2, and the matrix $\bar{\boldsymbol{Z}}_{21}$ repropagates the resulting contribution toward scatterer 1; finally, $\bar{\boldsymbol{Z}}_1^{-1}$ updates the currents values on scatterer 1. And so on for the subsequent terms $\boldsymbol{Y}_1^{(p)}$ for $p > 1$. Thus, the total currents $\sum_p \boldsymbol{Y}_1^{(p)}$ on scatterer 1 are due to the multiple scattering of the field inside the medium Ω_1.

4.3.2. *Numerical results*

4.3.2.1. *Coated circular cylinder*

Figure 4.6 shows the modulus of the total field in dB scale versus the abscissa x and the height z. Figure 4.7 shows the RCS in dB scale versus the scattering angle θ_{sca}. The radii of the two concentric circular cylinders are $a_1 = 3\lambda_0$, $a_2 = 2\lambda_0$, their centers are $C_1 = C_2 = (0, 0)$, the relative permittivities of media $\{\Omega_0, \Omega_1, \Omega_2\}$ are $\{\epsilon_{r0} = 1, \epsilon_{r1} = 2 + 0.1j, \epsilon_{r2} = 4 + 0.1j\}$, respectively. The incidence angle is $\theta_{\text{inc}} = 0$ and the polarization is TE. For the MoM, the number of samples per wavelength is $N_{\lambda_0} = 10$. In addition, the value of the following relative residual error defined as:

$$\epsilon_{\text{PILE}} = \frac{\text{norm}_{\theta_{\text{sca}}}(\text{RCS}_{\text{PILE}} - \text{RCS}_{\text{LU}})}{\text{norm}_{\theta_{\text{sca}}}(\text{RCS}_{\text{LU}})}, \qquad [4.33]$$

is given, where the “norm” stands for the norm two.

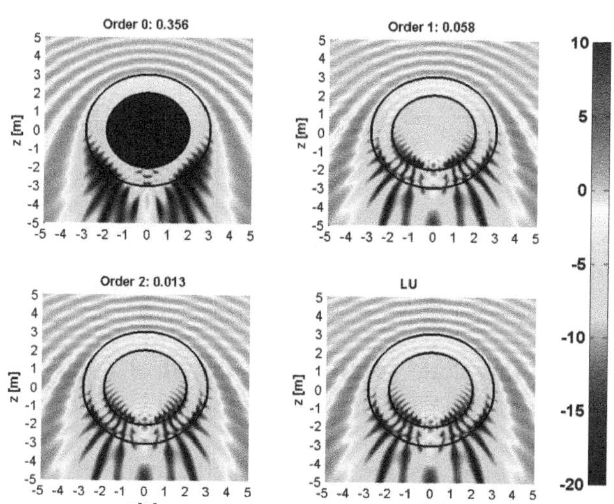

Figure 4.6. *Modulus of the total field in dB scale versus the abscissa x and the height z. The radii of the two concentric circular cylinders are $a_1 = 3\lambda_0$ and $a_2 = 2\lambda_0$, their centers are $C_1 = C_2 = (0, 0)$, the relative permittivities of media $\{\Omega_0, \Omega_1, \Omega_2\}$ are $\{\epsilon_{r0} = 1, \epsilon_{r1} = 2 + 0.1j, \epsilon_{r2} = 4 + 0.1j\}$, and $\lambda_0 = 1$ m. The incidence angle is $\theta_{inc} = 0$ and the polarization is TE*

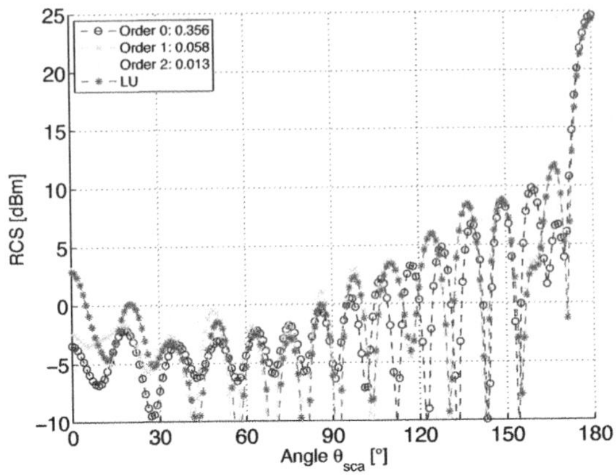

Figure 4.7. *RCS in dBm versus the scattering angle θ_{sca}. The parameters are the same as in Figure 4.6*

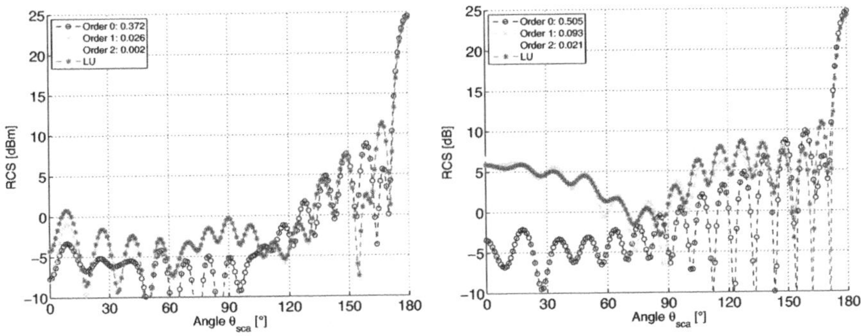

a) TM polarization

b) Scatterer 2 is PC

Figure 4.8. *Same results as in Figure 4.7, but for the a) TM polarization and b) scatterer 2 is perfectly conducting*

Figure 4.8 shows the same results as in Figure 4.7 but for the a) TM polarization and b) scatterer 2 is perfectly conducting.

For $P_{\text{PILE}} = 0$, in Figure 4.6, the field inside the cylinder 2 is zero because $\boldsymbol{X}_2 = 0$, whereas the field inside Ω_0 and Ω_1 differs from zero and corresponds to the field radiated by the cylinder 1. As the order P_{PILE} increases, Figures 4.7 and 4.8a) show that the RCS converges rapidly toward that obtained from a direct lower–upper (LU) inversion of the impedance matrix of the two cylinders. Indeed, since Ω_1 and Ω_2 are lossy media, the field inside the cylinders is damped. In addition, Figure 4.8b) shows that PILE also converges rapidly, which can be surprising, since P_{PILE} is related to the numbers of reflections between the two scatterers. A physical explanation is that, unlike a dielectric medium Ω_2, no field is transmitted into Ω_2 and then the upper part of cylinder 2 cannot interact with the lower part of the cylinder 1, which reduces the interaction between cylinders 1 and 2.

A means for knowing the convergence rate of PILE is to calculate the norm of the characteristic matrix, $\|\bar{M}_c\|$, which equals the modulus of its eigenvalue that has the highest modulus. For Figures 4.7 and 4.8a) and 4.8b), $\|\bar{M}_c\| = \{0.271, 0.105, 0.297\}$, respectively. For a large problem, the computation of this norm is very time consuming and it is not recommended.

The first advantage of PILE in comparison to a direct LU inversion is its ability to quantify the coupling between the two scatterers.

4.3.2.2. *Cylinder below a rough surface*

Figure 4.9 shows the modulus of the total field versus the abscissa x and the height z. Figure 4.10a) shows the normalized radar cross-section (NRCS) in dB scale versus the scattering angle θ_{sca}. The rough surface has a Gaussian surface height autocorrelation function of correlation length $L_c = \lambda_0$ and a standard deviation of height $\sigma_z = 0.5\lambda_0$. Its length is $L_1 = 40\lambda_0$ and the number of samples is $N_1 = 566$. The parameters of the perfectly conducting elliptical cylinder below the surface are $a_2 = 2\lambda_0$, $b_2 = \lambda_0$, $\alpha_2 = 0$, $C_2 = (0, -2)\lambda_0$ and $N_2 = 97$.

The relative permittivity of the medium Ω_1 is $\epsilon_{r1} = 2 + 0.1j$, the incidence angle is $\theta_{\text{inc}} = \pi/6$, the polarization is TE and $\lambda_0 = 1$ m. Figure 4.10b) shows the same results as in Figure 4.10a) but for the TM polarization.

Since Ω_1 is a lossy medium, PILE converges very rapidly, as shown in Figure 4.10. Figure 4.9 clearly shows the interference between the transmitted wave from Ω_0 into Ω_1 and the first reflected wave by the cylinder taken into account for $P_{\text{PILE}} = 1$. The comparison with the results computed from an LU inversion shows that it is not necessary to include the second reflection by the cylinder (it would correspond to $P_{\text{PILE}} = 2$). In addition, the shadow zone behind the cylinder is well predicted. For further simulations, see [BOU 08]. The authors showed that as σ_z and θ_{inc} increase and the polarization changes, the convergence order (order for which the convergence is reached) does not change and as the permittivity of medium Ω_1 increases, the convergence order decreases and as the major axis of the cylinder increases, it slightly increases. There are no simulations in which the order of convergence exceeds four. In general, PILE converges more rapidly than E-PILE because for the latter, the two scatterers are embedded in the lossless medium Ω_0, which is vacuum.

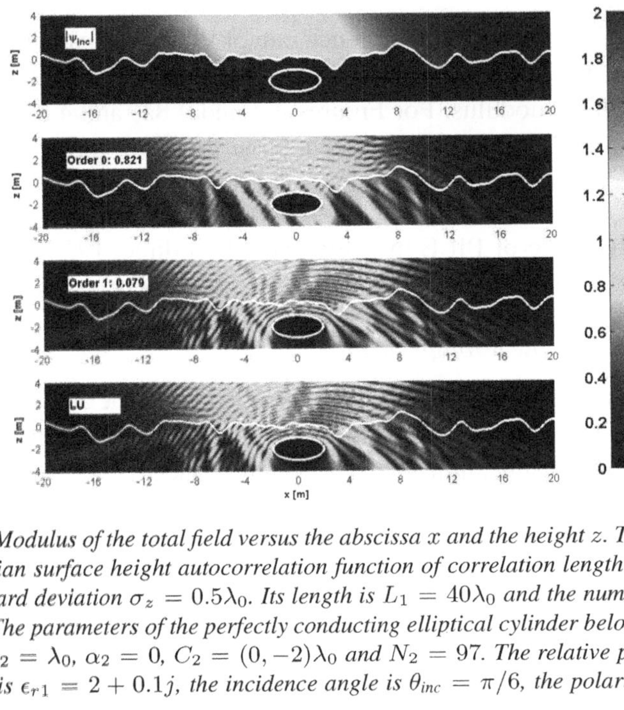

Figure 4.9. *Modulus of the total field versus the abscissa x and the height z. The rough surface has a Gaussian surface height autocorrelation function of correlation length $L_c = \lambda_0$ and of height standard deviation $\sigma_z = 0.5\lambda_0$. Its length is $L_1 = 40\lambda_0$ and the number of samples is $N_1 = 566$. The parameters of the perfectly conducting elliptical cylinder below the surface are $a_2 = 2\lambda_0$, $b_2 = \lambda_0$, $\alpha_2 = 0$, $C_2 = (0, -2)\lambda_0$ and $N_2 = 97$. The relative permittivity of the medium Ω_1 is $\epsilon_{r1} = 2 + 0.1j$, the incidence angle is $\theta_{inc} = \pi/6$, the polarization is TE and $\lambda_0 = 1$ m*

For Figures 4.10a) and 4.10b), the norm of the characteristic matrix is $\|\bar{M}_c\| = \{0.144, 0.137\}$, respectively.

4.3.2.3. *Random rough layer*

Figure 4.11 shows the modulus of the total field versus the abscissa x and the height z. The two random rough surfaces have Gaussian surface height autocorrelation functions. For the upper one: $L_{c1} = \lambda_0$, $\sigma_{z1} = 0.5\lambda_0$, $L_1 = 40\lambda_0$ and $N_1 = 800$. For the lower one: $L_{c2} = 2\lambda_0$, $\sigma_{z2} = 0.5\lambda_0$, $L_2 = 40\lambda_0$ and $N_2 = 560$. The thickness Ω_1 is $2\lambda_0$ ($C_2 = (0, -2\lambda_0)$), the relative permittivities of media $\Omega_{1,2}$ are $\epsilon_{r1,r2} = \{4 + 0.1j, 2 + 0.01j\}$, the incidence angle is $\theta_{\text{inc}} = \pi/6$, the polarization is TE and $\lambda_0 = 1$ m. Figure 4.12a) shows the corresponding NRCS and Figure 4.12b) shows the same results but for the TM polarization. As can be seen, PILE rapidly converges because Ω_1 and Ω_2 are lossy media.

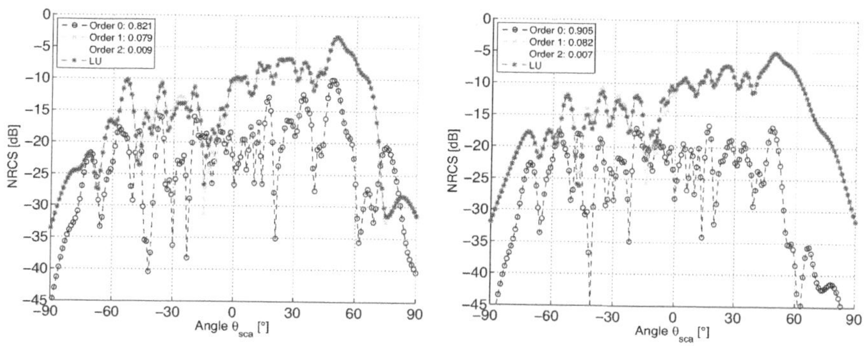

a) TE polarization b) TM polarization

Figure 4.10. *NRCS in dB scale versus the scattering angle θ_{sca}. Same parameters as in Figure 4.9*

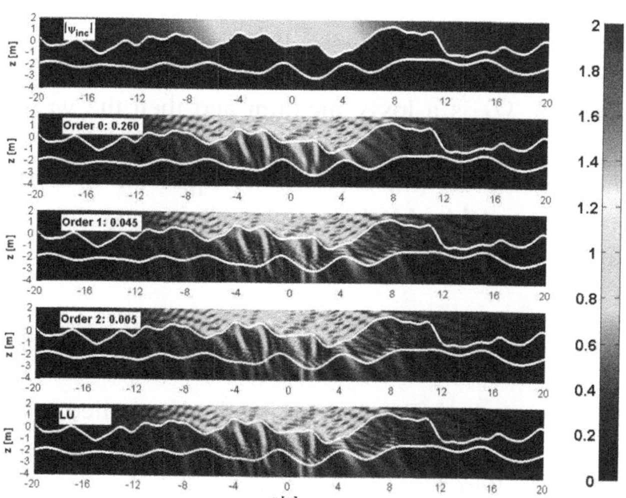

Figure 4.11. *Modulus of the total field versus the abscissa x and the height z. The two rough surfaces have Gaussian surface height autocorrelation functions. For the upper one: $L_{c1} = \lambda_0$, $\sigma_{z1} = 0.5\lambda_0$, $L_1 = 40\lambda_0$ and $N_1 = 800$. For the lower one: $L_{c2} = 2\lambda_0$, $\sigma_{z2} = 0.5\lambda_0$, $L_2 = 40\lambda_0$ and $N_2 = 560$. The thickness of Ω_1 is $2\lambda_0$ ($C_2 = (0, -2\lambda_0)$), the relative permittivities of the media $\Omega_{1,2}$ are $\epsilon_{r1,r2} = \{4 + 0.1j, 2 + 0.01j\}$, the incidence angle is $\theta_{inc} = \pi/6$, the polarization is TE and $\lambda_0 = 1$ m*

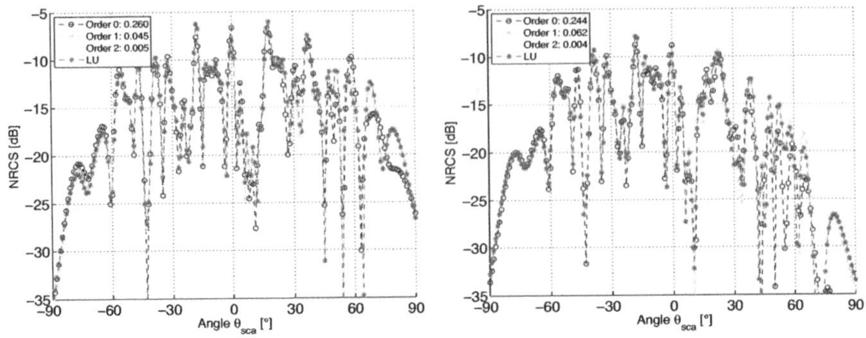

a) TE polarization b) TM polarization

Figure 4.12. *NRCS in dB scale versus the scattering angle θ_{sca}. Same parameters as in Figure 4.11*

For Figures 4.12a) and 4.12b), the norm of the characteristic matrix is $\|\bar{\boldsymbol{M}}_c\| = \{0.144, 0.137\}$, respectively.

4.3.2.4. *Conclusion*

The numerical results showed that PILE converges rapidly because its order is related to the multiple reflections between the two scatterers. For practical applications, Ω_1 is a lossy medium and then the wave is damped, which explains why PILE converges rapidly. If scatterer 2 is perfectly conducting, the PILE order increases in comparison to a dielectric Ω_2 medium. From the available MatLab codes, it is easy to verify this. In addition, for further simulations, the reader is referred to [DÉC 06, BOU 08].

4.4. PILE method combined with FB or PO

This section presents how PO and FB methods, applied to calculate the local interactions on a single scatterer, can be hybridized in the PILE algorithm to compute the field scattered by two scatterers.

4.4.1. *PILE hybridized with FB*

Equation [4.31] clearly shows that the calculation of $\bar{\boldsymbol{Z}}_i^{-1}\boldsymbol{u}$ is required, in which $\bar{\boldsymbol{Z}}_i$ is the impedance matrix of scatterer i in free space. Then, to calculate the local interactions on a random rough surface, the FB method can be applied. For any scatterer below a rough surface and for a rough layer, Bourlier *et al.* [BOU 08] and Déchamps and Bourlier [DÉC 07a, DÉC 07b] showed that the order P_{FB} of convergence of FB is obtained by considering only scattering from the single rough surface (without the scatterer). This can be explained by the fact that the inversion of the impedance matrix is independent of the incident field $\boldsymbol{u}$.

For the example shown in Figure 4.11, first, the FB order for the lower and upper rough surfaces must be determined. Then, each surface is considered to be alone, and for a residual relative error smaller than 0.01, we find $\{p_{\text{FB},1} = 9, p_{\text{FB},2} = 8\}$ for the TE polarization and $\{p_{\text{FB},1} = 7, p_{\text{FB},2} = 7\}$ for the TM polarization. Combining PILE with FB for the calculation of the local interactions on each surface, the NRCSs depicted in Figure 4.13 are computed from PILE+FB+FB. In comparison with Figure 4.12, we can see that the residual errors (defined from equation [4.33]) are nearly identical, meaning that FB has converged for the calculation of the local interactions on each scatterer.

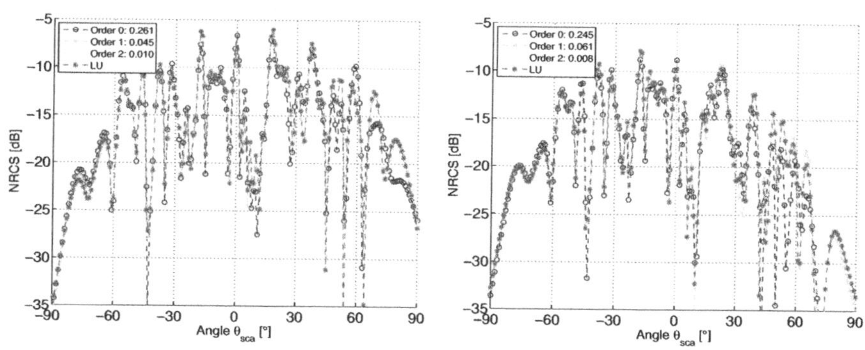

a) TE polarization b) TM polarization

Figure 4.13. *NRCS in dB scale versus the scattering angle θ_{sca} computed from PILE+FB+FB. Same parameters as in Figure 4.12*

It is also possible to apply the FB only on scatterer 1 or 2 and to apply an LU inversion on the other scatterer. For an object below a rough surface, the FB is applied onto the surface, whereas an LU inversion is applied onto the object, since FB does not converge for a closed surface. With the same parameters as in Figure 4.9, Figure 4.14 shows the NRCS computed from PILE+FB+LU with

$\{p_{\text{FB},1} = 9, 7\}$ for the TE and TM polarizations, respectively. As was the case previously, in comparison to Figure 4.10, we can see that the residual errors are quasi-identical, meaning that the FB has converged for the calculation of the local interactions on the rough surface.

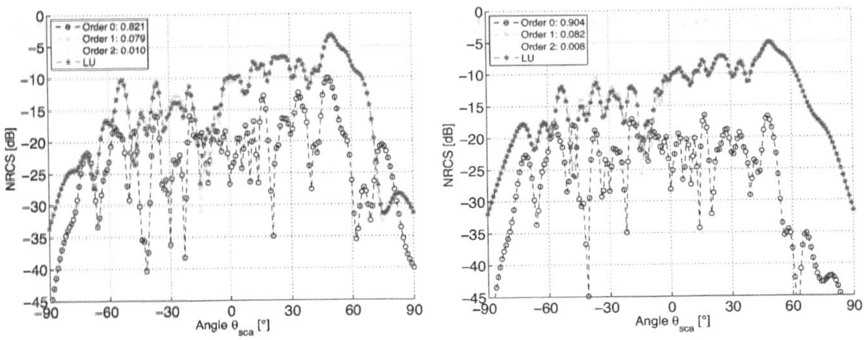

a) TE polarization b) TM polarization

Figure 4.14. *NRCS in dB scale versus the scattering angle θ_{sca} computed from PILE+FB+LU. Same parameters as in Figure 4.10*

4.4.2. *PILE hybridized with PO*

This section focuses on the integration of PO in the PILE algorithm. For E-PILE, this hybridization is thoroughly explained in section 3.4.2 by considering the scattering by two illuminated perfectly conducting scatterers. Here, the problem is slightly different and more complicated, because the illuminated scatterer must be dielectric to interact with the non-illuminated scatterer.

For a coated cylinder illuminated by a plane wave of incidence angle $\theta_{\text{inc}} = 0$, if the PO approximation is applied at first order (only point-to-point local interactions are considered, i.e. no multiple reflections) on the illuminated scatterer (number 1), then in the shadow zone (lower part of the cylinder), the currents vanish. Thus, the PO at first order can predict bad results. To overcome this issue, the field transmitted into the cylinder must be computed by iterating the PO approximation on each scatterer. For an object below a rough surface, the problem is similar. For a rough layer, the inverse of the impedance matrix of a single dielectric rough surface must be calculated from the PO. This operation is not trivial and it is a prospect considered in this book. In conclusion, to simplify the formulation, we assume that the non-illuminated scatterer is perfectly conducting and the PO is applied only onto this scatterer.

4.4.2.1. *Formulation*

For scatterer 2, the total field on the object surface in the medium Ω_1 due to a single reflection is given under the PO approximation by equation [2.62]:

$$\psi_2(\boldsymbol{r}) = 2\begin{cases} \psi_{\text{inc},1}(\boldsymbol{r}) & \boldsymbol{r} \in S_{2,\text{Ill}} \\ 0 & \boldsymbol{r} \in S_{2,\text{Sha}} \end{cases} \text{ and } \frac{\partial \psi_2(\boldsymbol{r})}{\partial n} = 0 \; \forall \; \boldsymbol{r} \in S_2, \text{ TM polarization,} \tag{4.34}$$

$$\frac{\partial \psi_2(\boldsymbol{r})}{\partial n} = 2\begin{cases} \dfrac{\partial \psi_{\text{inc},1}(\boldsymbol{r})}{\partial n} & \boldsymbol{r} \in S_{2,\text{Ill}} \\ 0 & \boldsymbol{r} \in S_{2,\text{Sha}} \end{cases} \text{ and } \psi_2(\boldsymbol{r}) = 0 \; \forall \; \boldsymbol{r} \in S_2, \text{ TE polarization,} \tag{4.35}$$

where $\psi_{\text{inc},1}$ is the field inside Ω_1 radiated from the currents on scatterer 1, $S_{2,\text{Ill}}$ is the illuminated surface and $S_{2,\text{Sha}}$ is the shadowed surface ($S_2 = S_{2,\text{Ill}} \cup S_{2,\text{Sha}}$). Then, under the PO approximation, the inverse of the matrix impedance $\bar{\boldsymbol{Z}}_2^{-1}$ is a diagonal matrix of elements equal to 2 for the TM polarization and elements $2\partial(\bullet)/\partial n$ for the TE polarization. The complexity of the inversion is then $\mathcal{O}(1)$ instead of $\mathcal{O}(N_2^3)$ from a direct LU inversion.

For example, for the calculation of $\boldsymbol{Y}_1^{(1)} = \bar{\boldsymbol{Z}}_1^{-1}\bar{\boldsymbol{Z}}_{21}\bar{\boldsymbol{Z}}_2^{-1}\bar{\boldsymbol{Z}}_{12}\boldsymbol{Y}_1^{(0)}$, where $\boldsymbol{Y}_1^{(0)} = \bar{\boldsymbol{Z}}_1^{-1}\boldsymbol{b}_1$, first, the vector $\boldsymbol{Y}_1^{(0)}$ is multiplied by the matrix $\bar{\boldsymbol{Z}}_{12}$ giving $\boldsymbol{u} = \bar{\boldsymbol{Z}}_{12}\boldsymbol{Y}_1^{(0)}$. It can be considered as an incident field for scatterer 2. If the PO is applied onto scatterer 2, then some elements of $\bar{\boldsymbol{Z}}_{12}$ can be zero due to the fact that a point on scatterer 2 cannot be viewed from a point of scatterer 1. For a convex object, this condition is satisfied if $\hat{\boldsymbol{n}}_2 \cdot (\boldsymbol{r}_2 - \boldsymbol{r}_1) > 0$. Then, the elements of the modified matrix $\bar{\boldsymbol{Z}}'_{12}$ are:

$$\begin{aligned} Z'_{12,mn} &= Z_{12,mn}\frac{1 - \text{sgn}\left[(\boldsymbol{r}_{2,m} - \boldsymbol{r}_{1,n}) \cdot \hat{\boldsymbol{n}}_{2,m}\right]}{2} \\ &= Z_{12,mn}\frac{1 + \text{sgn}\left[(x_{2,m} - x_{1,n})v_{2,m}\gamma_{2,m} - (z_{2,m} - z_{1,n})v_{2,m}\right]}{2}. \end{aligned} \tag{4.36}$$

Next, $\boldsymbol{u} = \bar{\boldsymbol{Z}}_{12}\boldsymbol{Y}_1^{(0)}$ is multiplied by $\bar{\boldsymbol{Z}}_2^{-1}$. Then, if the PO is applied onto scatterer 2, currents $\boldsymbol{v} = \bar{\boldsymbol{Z}}_2^{-1}\boldsymbol{u}$ are computed as follows:

$$\begin{aligned}\boldsymbol{v} = \bar{\boldsymbol{Z}}_2^{-1}\boldsymbol{u} = \bar{\boldsymbol{Z}}_2^{-1}\bar{\boldsymbol{Z}}'_{12}\boldsymbol{Y}_1^{(0)} &= \bar{\boldsymbol{D}}\begin{bmatrix}\bar{\boldsymbol{A}}'_{12} & \frac{1}{\rho_{01}}\bar{\boldsymbol{B}}'_{12}\end{bmatrix}\begin{bmatrix}\boldsymbol{w}_1 \\ \boldsymbol{w}_2\end{bmatrix} \\ &= \bar{\boldsymbol{D}}\left[\bar{\boldsymbol{A}}'_{12}\boldsymbol{w}_1 + \frac{1}{\rho_{01}}\bar{\boldsymbol{B}}'_{12}\boldsymbol{w}_2\right],\end{aligned} \quad [4.37]$$

where $\bar{\boldsymbol{D}}$ is the diagonal matrix of elements equal to 2 at point $\boldsymbol{r}_{2,n}$ for the TM polarization and elements equal to $2\partial \bullet /\partial n$ for the TE polarization. In addition, $\boldsymbol{Y}_1^{(0)} = [\boldsymbol{w}_1^{\mathrm{T}}\ \boldsymbol{w}_2^{\mathrm{T}}]^{\mathrm{T}}$ ($\boldsymbol{w}_1$ and $\boldsymbol{w}_2$ are vectors of length N_1). For the TM polarization, the above equation needs to compute $\partial\bar{\boldsymbol{A}}_{12}/\partial n$ and $\partial\bar{\boldsymbol{B}}_{12}/\partial n$ at point $\boldsymbol{r}_2$. From equation [4.20], we then show that

$$\left.\frac{\partial B_{12,mn}}{\partial n}\right|_{\boldsymbol{r}_{2,m}} = \frac{jk_1 v_{2,m}|\Delta_{1,n}|\sqrt{1+\gamma_{1,n}^2}\mathrm{H}_{11}}{4r_{12}\sqrt{1+\gamma_{2,m}^2}}\left(z_{12} - \gamma_{2,m}x_{12}\right), \quad [4.38]$$

and

$$\left.\frac{\partial A_{12,mn}}{\partial n}\right|_{\boldsymbol{r}_{2,m}} = -\frac{jk_1 v_{1,n}|\Delta_{1,n}|v_{2,m}}{4\sqrt{1+\gamma_{2,m}^2}}\left[w_{00} + w_{10}\left(\gamma_{1,n}+\gamma_{2,m}\right) + w_{11}\gamma_{1,n}\gamma_{2,m}\right], \quad [4.39]$$

where

$$\begin{cases} w_{00} = \dfrac{k_1 z_{12}^2 \mathrm{H}_{10}}{r_{12}^2} + \dfrac{(x_{12}^2 - z_{12}^2)\mathrm{H}_{11}}{r_{12}^3} \\ w_{10} = \dfrac{x_{12}z_{12}}{r_{12}^3}\left(2\mathrm{H}_{11} - \mathrm{H}_{10}k_1 r_{12}\right) \\ w_{11} = \dfrac{k_1 x_{12}^2 \mathrm{H}_{10}}{r_{12}^2} + \dfrac{(z_{12}^2 - x_{12}^2)\mathrm{H}_{11}}{r_{12}^3} \end{cases}, \quad [4.40]$$

and $x_{12} = x_{1,n} - x_{2,m}$, $z_{12} = z_{1,n} - z_{2,m}$, $r_{12} = \sqrt{x_{12}^2 + z_{12}^2}$, $\mathrm{H}_{10} = \mathrm{H}_0^{(1)}(k_1 r_{12})$, $\mathrm{H}_{11} = \mathrm{H}_1^{(1)}(k_1 r_{12})$.

4.4.2.2. *Numerical results: coated elliptical cylinder*

Figure 4.16 shows the RCS in dBm versus the angle θ_{sca}, respectively, and for both TE and TM polarizations. The parameters are given in Figure 4.15a). In the figure, the label "PILE+LU+PO" means that the PILE is hybridized with LU for the calculation of the local interactions on cylinder 1 and with PO for the calculation of the local interactions on cylinder 2. In addition, the value of the following residual relative error defined as

$$\epsilon_{\text{PILE}} = \frac{\text{norm}_{\theta_{\text{sca}}}\left(\text{RCS}_{\text{PILE+LU+PO}} - \text{RCS}_{\text{LU}}\right)}{\text{norm}_{\theta_{\text{sca}}}\left(\text{RCS}_{\text{LU}}\right)}, \qquad [4.41]$$

is given in parentheses.

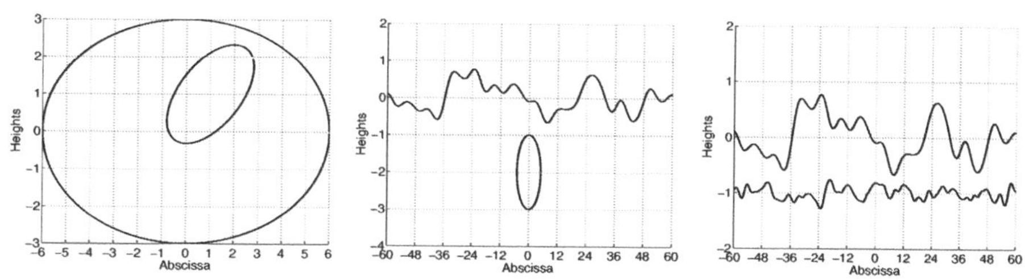

Figure 4.15. *a) A coated elliptical cylinder: semi-major axes* $a_1 = 6\lambda_0$, $a_2 = 2\lambda_0$; *semi-minor axes* $b_1 = 3\lambda_0$, $b_2 = \lambda_0$; *centers* $C_1 = (0,0)$, $C_2 = (1,1)\lambda_0$; *rotation angles* $\alpha_1 = 0$, $\alpha_2 = 30^o$ *and incidence angle* $\theta_{inc} = 30^o$. *b) Elliptical cylinder below a rough surface:* $a_2 = 5\lambda_0$, $b_2 = 2\lambda_0$, $C_2 = (0,-2)\lambda_0$, $\alpha_2 = 0$, $L_1 = 120\lambda_0$, $\sigma_z = 0.5\lambda_0$, $L_c = 5\lambda_0$, *the surface height autocorrelation function is Gaussian, the parameter of the Thorsos wave (defined in equation* [1.32]*) is* $g = L_2/6$ *and incidence angle* $\theta_{inc} = 30^o$. *c) Rough layer:* $L_1 = L_2 = 120\lambda_0$, $\sigma_{z1} = 0.5\lambda_0$, $\sigma_{z2} = 0.1\lambda_0$, $L_{c1} = 5\lambda_0$, $L_{c2} = 2\lambda_0$, *the surface height autocorrelation function for both surfaces is Gaussian, the parameter of the Thorsos wave (defined from equation* [1.32]*) is* $g = L_2/6$ *and the incidence angle* $\theta_{inc} = 30^o$

As the PILE order increases, the results converge toward those obtained from a direct LU inversion of the impedance matrix of the two scatterers. In addition, the results obtained from "PILE+LU+PO" match well with those obtained without hybridization (PILE) and the differences have a minor impact on the RCS. This hybridization is therefore very efficient and allows us to reduce the complexity to $\mathcal{O}(N_1^3)$ (LU inversion on scatterer 1) instead of $\mathcal{O}((N_1 + N_2)^3)$ for a direct LU inversion.

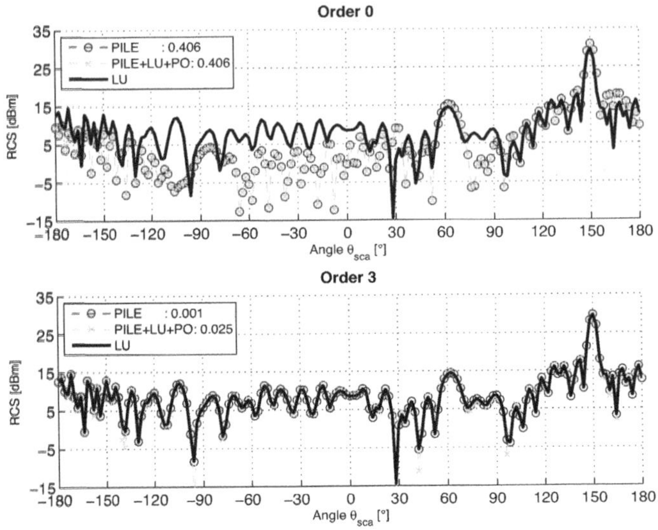

a) RCS and TE

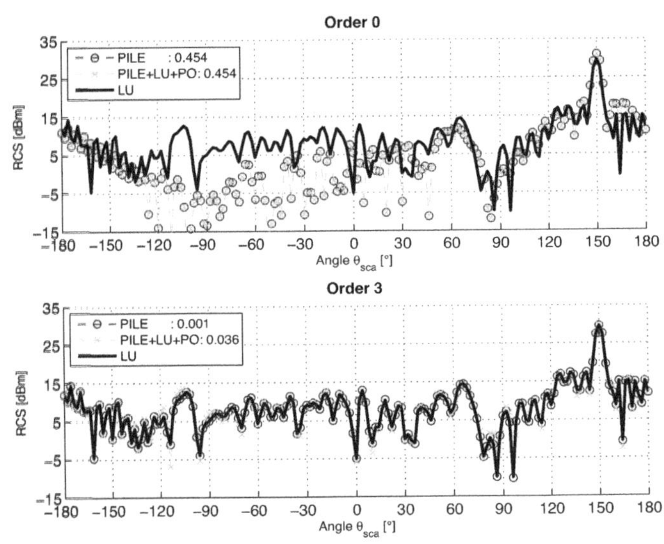

b) RCS and TM

Figure 4.16. *RCS in dBm versus θ_{sca}. a) TE polarization. b) TM polarization. The parameters are given in Figure 4.15a)*

4.4.2.3. *Numerical results: object below a random rough surface*

Figure 4.17 shows the NRCS in dB versus the scattering angle θ_{sca}, and for both the TE and TM polarizations. The parameters are given in Figure 4.15b). In the figure, the label "PILE+FB+PO" means that PILE is hybridized with FB for the calculation of the local interactions on the rough surface and with PO for the calculation of the local interactions on the cylinder. The order of FB is 6. As we saw previously, as the PILE order increases, the results converge toward those obtained from a direct LU inversion of the impedance matrix of the two scatterers. In addition, the results computed from "PILE+FB+PO" match well with those obtained without hybridization ("PILE") and the differences have a minor impact on the RCS. This hybridization is then very efficient and allows us to reduce the complexity to $\mathcal{O}(N_1^2)$ (FB on scatterer 1) instead of $\mathcal{O}((N_1 + N_2)^3)$ for a direct LU inversion.

4.4.2.4. *Numerical results: random rough layer*

Figure 4.18 shows the NRCS in dB versus the scattering angle θ_{sca}, respectively, for the TE polarization and the TM polarization. The parameters are given in Figure 4.15c). A very good agreement between the different methods is obtained. The zero order of the PILE method corresponds to the scattering from scatterer 1 alone.

For a rough layer, the FB can be applied on both surfaces and can also be accelerated by hybridizing the SA. The resulting complexity is then $\mathcal{O}(N_1 + N_2 + N_1 N_2)$, in which $\mathcal{O}(N_1 N_2)$ corresponds to the complexity of the matrix-vector products. The SA can also be applied to reduce this complexity to $\mathcal{O}(N_1) + \mathcal{O}(N_2)$ leading to a very efficient method. For more details, see [DÉC 07b]. Similarly, for slightly rough surfaces (where typically the standard height deviation does not exceed the wavelength), the BMIA/CAG can be applied to reduce both the complexity of the calculations of the local interactions and the matrix-vector products [DÉC 07a].

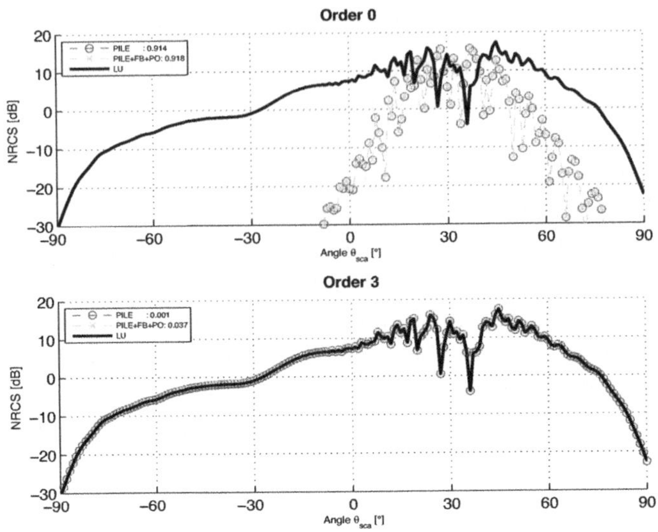

a) NRCS and TE.

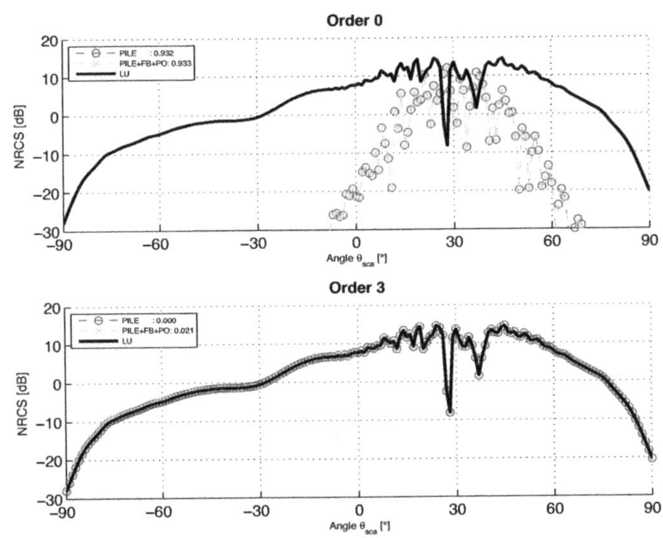

b) NRCS and TM.

Figure 4.17. *NRCS in dB versus the scattering angle θ_{sca}. a) TE polarization. b) TM polarization. The simulation parameters are given in Figure 4.15b*

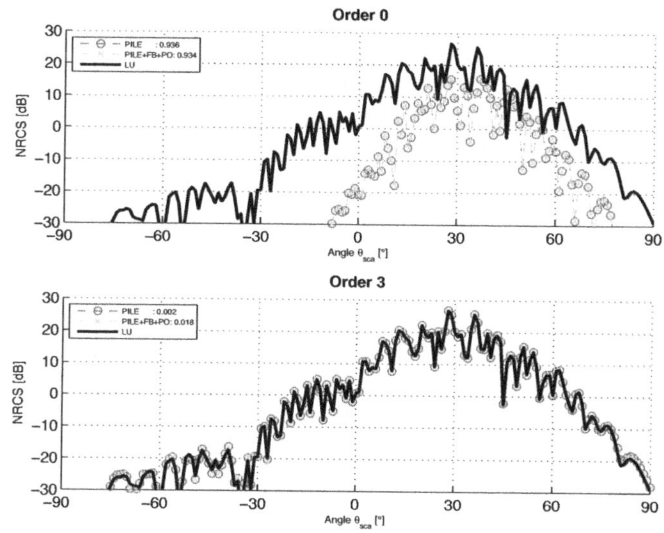

a) NRCS and TE

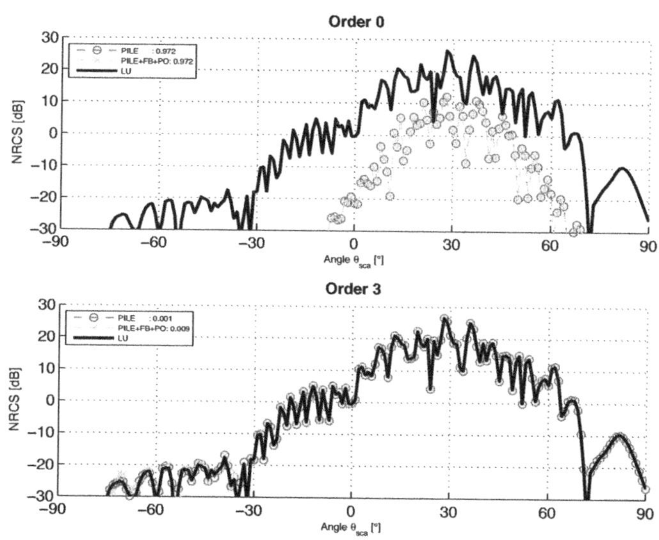

b) NRCS and TM

Figure 4.18. *NRCS in dB versus θ_{sca}. a) TE polarization. b) TM polarization. The simulation parameters are given in Figure 4.15c*

4.5. Conclusion

This chapter presents an efficient numerical method to calculate the field scattered by two scatterers where only one is illuminated. The PILE method starts from the MoM and the impedance matrix of the two scatterers is then inverted by blocks from the Taylor series expansion of the inverse of the Schur complement. Its main use is that it is rigorous, with a simple formulation and has a straightforward physical interpretation. Actually, this last property relies on the fact that each block of the impedance matrix is linked to a particular and quasi-independent physical process occurring during the multiple scattering process between the two scatterers: local interactions on each interface, corresponding to the inversion of the impedance matrix of each scatterer in free space, and both upward and downward coupling. Furthermore, the PILE method allows us to use any fast method developed for a single interface. Here, for a scatterer 2 that is assumed to conduct perfectly, in order to decrease the complexity of PILE, PO and/or FB have been hybridized with PILE and according to the scenario, this hybridization gives satisfactory results.

Appendix

MatLab Codes

This appendix presents the codes developed using the MatLab software for presenting the numerical results.

A program, whose name begins by "A_", is a main program, which generates the figures. A program, whose name begins by "F_", is a function, which can be called from the main program.

At the beginning of each main program, the called functions are mentioned as comments.

A1.1. Chapter 1

Table A1.1 presents the programs used in Chapter 1.

Name	Corresponding figures
A_Surface_Generation.m	1.4, 1.5
A_Surface_Generation_Spectra.m	1.6, 1.7

Table A1.1. *Relation between figures of Chapter 1 and names of the programs*

A1.2. Chapter 2

Table A1.2 presents the programs used in Chapter 2.

Name	Corresponding figures
A_Cylinder_Ana_DI_PC	2.3–2.6
A_Cylinder_Ana_MoM_PC.m	2.7–2.9
A_Cylinder_Ana_MoM_DI.m	2.10a), 2.10b), 2.11
A_CylinderE_MoM_DI_IBC_PC.m	2.12, 2.13
A_Plate_MoM_PO_PC.m	2.16–2.18
A_CylinderE_MoM_PO_PC.m	2.20, 2.21
A_Surface_MoM_PO_DI_IBC_PC.m	2.22, 2.23, 2.25, 2.26, 2.27
A_Surface_MoM_FB_DI_IBC_PC.m	2.28, 2.29

Table A1.2. *Relation between figures of Chapter 2 and names of the programs*

A1.3. Chapter 3

Table A1.3 presents the programs used in Chapter 3.

Name	Corresponding figures
A_2CylindersE_LU.m	3.2–3.5
A_NScatterers_LU_PC	3.6
A_2Scatterers_EPILE_FB_PC	3.8–3.11
A_2CylindersE_EPILE	3.12, 3.13
A_2Scatterers_EPILE_LU_FB_DI	3.14, 3.15
A_2Scatterers_EPILE_LU_FB_PO_PC	3.16–3.23

Table A1.3. *Relation between figures of Chapter 3 and names of the programs*

A1.4. Chapter 4

Table A1.4 presents the programs used in Chapter 4.

Name	Corresponding figures
A_CoatedCylinder_LU_Ana	4.2, 4.3
A_CoatedCylinderE_LU	4.4, 4.5
A_CoatedCylinderE_PILE.m	4.6–4.8
A_SurfaceObject_PILE_FB_LU.m	4.9, 4.10
A_Layer_PILE_FB_FB_LU.m	4.11–4.13
A_CoatedCylinderE_PILE_PO.m	4.15a), 4.16
A_SurfaceObjectLayer_PILE_PO_PC.m	4.15b), 4.15c), 4.17, 4.18

Table A1.4. *Relation between figures of Chapter 4 and names of the programs*

Bibliography

[ABR 70] ABRAMOWITZ M., STEGUN I.A., *Handbook of Mathematical Functions*, Dover Publications, Inc., New York, 1970.

[ADA 96] ADAMS R.J., BROWN G.S., "An iterative solution of one-dimensional rough surface scattering problems based on a factorization of the Helmholtz operator", *IEEE Transactions on Antennas and Propagation*, vol. 47, no. 4, pp. 765–767, 1996.

[AHM 08] AHMED S., NAQVI Q.A., "Electromagnetic scattering from a perfect electromagnetic conductor cylinder buried in a dielectric half-space", *Progress in Electromagnetics Research*, vol. 78, pp. 25–38, 2008.

[BOU 99] BOURLIER C., Rayonnement infrarouge d'une surface stochastique – Application au domaine océanique, PhD Dissertation, University of Nantes, France, 1999.

[BOU 00] BOURLIER C., SAILLARD J., BERGINC G., "Intrinsic infrared radiation of the sea surface", *Progress in Electromagnetic Research*, vol. 27, pp. 185–335, 2000.

[BOU 04a] BOURLIER C., BERGINC G., "Multiple scattering in the high-frequency limit with second-order shadowing function from 2-D anisotropic rough dielectric surfaces: I. theoretical study", *Waves in Random Complex Media*, vol. 14, no. 3, pp. 229–252, 2004.

[BOU 04b] BOURLIER C., BERGINC G., "Multiple scattering in the high-frequency limit with second-order shadowing function from 2D anisotropic rough dielectric surfaces: II. comparison with numerical results", *Waves in Random Complex Media*, vol. 14, no. 3, pp. 253–276, 2004.

[BOU 05] BOURLIER C., DÉCHAMPS N., PINEL N., "Comparison of asymptotic backscattering models (SSA, WCA and LCA) from one-dimensional ocean-like surfaces", *IEEE Transactions on Antennas and Propagation*, vol. 53, no. 5, pp. 1640–1652, 2005.

[BOU 08] BOURLIER C., KUBICKÉ G., DÉCHAMPS N., "A fast method to compute scattering by a buried object under a randomly rough surface: PILE combined to FB-SA", *Journal of the Optical Society of America A*, vol. 25, no. 4, pp. 891–902, 2008.

[BOU 11a] BOURLIER C., KUBICKÉ G., "HF ground wave propagation over a curved rough sea surface", *Waves in Random and Complex Media*, vol. 21, no. 1, pp. 23–43, 2011.

[BOU 11b] BOURLIER C., KUBICKÉ G., "HF ground wave propagation over a curved rough sea surface in the presence of islands", *Waves in Random and Complex Media*, vol. 21, no. 3, pp. 469–484, 2011.

[BOW 87] BOWMAN J.L., SENIOR T.B.A., USLENGHI P. L.E., *Electromagnetic and Acoustic Scattering by Simple Shapes*, Hemisphere Publishing Corporation, New York, 1987.

[BRE 80] BREKHOVSKIKH L.M., *Waves in Layered Media, 2nd ed.*, Academic Press, New York, 1980.

[CHO 00] CHOU H.T., JOHNSON J.T., "Formulation of the forward-backward method using novel spectra acceleration for the modeling of scattering from impedance rough surfaces", *IEEE Transactions on Geoscience and Remote Sensing*, vol. 38, no. 1, pp. 605–607, 2000.

[CHO 02] CHOU H.T., JOHNSON J.T., "A novel acceleration algorithm for the computation of scattering from rough surfaces with the forward-backward method", *Radio Science*, vol. 33, pp. 1277–1287, 2002.

[DÉC 06] DÉCHAMPS N., DE BEAUCOUDREY N., BOURLIER C., *et al.*, "Fast numerical method for electromagnetic scattering by rough layered interfaces: propagation-inside-layer expansion method", *Journal of the Optical Society of America A*, vol. 23, no. 2, pp. 359–369, 2006.

[DÉC 07a] DÉCHAMPS N., BOURLIER C., "Electromagnetic scattering from a rough layer: propagation-inside-layer expansion method combined to an updated BMIA/CAG approach", *IEEE Transactions on Antennas and Propagation*, vol. 55, no. 10, pp. 2790–2802, 2007.

[DÉC 07b] DÉCHAMPS N., BOURLIER C., "Electromagnetic scattering from a rough layer: propagation-inside-layer expansion method combined to the forward-backward novel spectral acceleration", *IEEE Transactions on Antennas and Propagation*, vol. 55, no. 12, pp. 3576–3586, 2007.

[DON 07] DONG C., WANG C., WEI X., *et al.*, "EM scattering from complex targets above a slightly rough surface", *Progress in Electromagnetic Research*, vol. 3, no. 5, pp. 685–688, 2007.

[ELF 97] ELFOUHAILY T., CHAPRON B., KATSAROS K., *et al.*, "A unified directional spectrum for long and short wind-driven waves", *Journal of Geophysical Research*, vol. 102, no. C7, pp. 781–796, 1997.

[ELF 04] ELFOUHAILY T., GUÉRIN C.-A., "Critical survey of approximate scattering wave theories from random rough surfaces (topical review)", *Waves in Random Complex Media*, vol. 14, no. 4, pp. R1–R40, 2004.

[FIA 12] FIAZ A.F., FREZZA F., PAJEWSKI L., *et al.*, "Scattering by a circular cylinder buried under a slightly rough surface: the cylindrical-wave approach", *IEEE Transactions on Antennas and Propagation*, vol. 60, no. 6, pp. 2834–2842, 2012.

[HAR 68] HARRINGTON R.F., *Field Computation by Moment Method*, Macmillan, New York, 1968.

[HEN 07] HENIN B.H., ELSHERBENI A.Z., SHARKAWY M. H.A., "Oblique incidence plane wave scattering from an array of circular dielectric cylinders", *Progress in Electromagnetics Research*, vol. 68, pp. 261–279, 2007.

[HOL 98] HOLLIDAY D., DERAAD JR. L.L., ST-CYR G.J., "Forward-backward method for scattering from imperfect conductors", *IEEE Transactions on Antennas and Propagation*, vol. 46, no. 1, pp. 101–107, 1998.

[IOD 02] IODICE A., "Forward-backward method for scattering from dielectric rough surfaces", *IEEE Transactions on Antennas and Propagation*, vol. 50, no. 7, pp. 901–911, 2002.

[ISH 86] ISHIMARU A., LE C., KUGA K., *et al.*, "Polarimetric scattering theory for high slope rough surface", *Progress in Electromagnetic Research*, vol. 14, pp. 1–36, 1986.

[JOH 02] JOHNSON J.T., "A numerical study of scattering from an object above a rough surface", *IEEE Transactions on Antennas and Propagation*, vol. 50, no. 10, pp. 1361–1367, 2002.

[KAP 96] KAPP D.A., BROWN G.S., "A new numerical method for rough-surface scattering calculations", *IEEE Transactions on Antennas and Propagation*, vol. 44, no. 5, pp. 711–722, 1996.

[KON 05] KONG J.A., *Electromagnetic Wave Theory*, EMW Publishing, Cambridge, MA, 2005.

[KUB 08] KUBICKÉ G., BOURLIER C., SAILLARD J., "Scattering by an object above a randomly rough surface from a fast numerical method: extended PILE method combined to FB-SA", *IEEE Transactions on Antennas and Propagation*, vol. 18, no. 3, pp. 495–519, 2008.

[KUB 10a] KUBICKÉ G., BOURLIER C., "A fast hybrid method for scattering from a large object with dihedral effects above a large rough surface", *IEEE Transactions on Antennas and Propagation*, vol. 59, no. 1, pp. 189–198, 2010.

[KUB 10b] KUBICKÉ G., BOURLIER C., SAILLARD J., "Monostatic radar cross section of an object above a sea surface from a rigorous method", *Comptes-rendus Physiques de l'Académie des Sciences*, vol. 11, no. 1, pp. 68–75, 2010.

[KUB 10c] KUBICKÉ G., BOURLIER C., SAILLARD J., "Scattering from canonical objects above a sea-like 1D rough surface from a rigorous fast method", *Waves in Random and Complex Media*, vol. 20, no. 1, pp. 156–178, 2010.

[KUB 11] KUBICKÉ G., YAHIA Y.A., BOURLIER C., *et al.*, "Bridging the gap between the babinet principle and the physical optics approximation: scalar problem", *IEEE Transactions on Antennas and Propagation*, vol. 49, no. 12, pp. 4725–4732, 2011.

[KUN 91] KUNT M., *Techniques modernes de traitement numériques des signaux*, Presses polytechniques universitaires romandes, Lausanne, Switzerland, 1991.

[LEE 11] LEE S.-C., "Scattering by closely spaced parallel non homogeneous cylinders in an absorbing medium", *Journal of the Optical Society of America A*, vol. 28, no. 9, pp. 1812–1819, 2011.

[LI 11] LI H., PINEL N., BOURLIER C., "A monostatic illumination function with surface reflections from one-dimensional rough surfaces", *Waves in Random Complex Media*, vol. 21, no. 1, pp. 105–134, 2011.

[LIU 04] LIU P., JIN Y.Q., "The finite-element method with domain decomposition for electromagnetic bistatic scattering from the comprehensive model of a ship on and a target above a large scale rough sea surface", *IEEE Transactions on Geoscience and Remote Sensing*, vol. 42, no. 5, pp. 950–956, 2004.

[MAD 95] MADRAZO A., NIETO-VESPERINAS M., "Scattering of electromagnetic waves from a cylinder in front of a conducting plane", *Journal of the Optical Society of America A*, vol. 12, pp. 1298–1309, 1995.

[OGI 91] OGILVY J.A., *Theory of Wave Scattering from Random Rough Surfaces*, Institute of Physics, Bristol, 1991.

[PAW 11] PAWLIUK P., YEDLIN M., "Scattering from cylinders using the two-dimensional vector plane wave spectrum", *Journal of the Optical Society of America A*, vol. 28, no. 6, pp. 1177–1184, 2011.

[PAW 12] PAWLIUK P., YEDLIN M., "Scattering from cylinders near a dielectric half-space using a general method of images", *Journal of the Optical Society of America A*, vol. 66, no. 11, pp. 5296–5304, 2012.

[PIN 99] PINO M.R., LANDESA L., RODRIGUEZ J.L., *et al.*, "The generalized forward-backward method for analyzing the scattering from targets on ocean-like rough surfaces", *IEEE Transactions on Antennas and Propagation*, vol. 47, no. 6, pp. 961–969, 1999.

[PIN 13] PINEL N., BOURLIER C., *Electromagnetic Wave Scattering from Random Rough Surfaces: Asymptotic Models*, ISTE, London, John Wiley & Sons, New York, 2013.

[PRE 92] PRESS W.H., TEUTOLSKY S.A., VETTERLING W.T., *et al.*, *Numerical Recipies, 2nd ed.*, Cambridge University Press, Cambridge, 1992.

[SAI 01] SAILLARD M., SENTENAC A., "Rigorous solutions for electromagnetic scattering from rough surfaces (topical review)", *Waves in Random Complex Media*, vol. 11, no. 3, pp. R103–R137, 2001.

[SOM 64] SOMMERFELD A., *Lectures on Theoretical Physics*, vol. 6, Academic Press, New York, 1964.

[THO 88] THORSOS E.I., "The validity of the Kirchhoff approximation for rough surface scattering using a Gaussian roughness spectrum", *Journal of the Acoustical Society of America*, vol. 86, no. 1, pp. 78–92, 1988.

[TOR 00] TORRUNGRUENG D., CHOU H.T., JOHNSON J.T., "A novel acceleration algorithm for the computation of scattering from two-dimensional large-scale perfectly conducting random rough surfaces with the forward-backward method", *IEEE Transactions on Geoscience and Remote Sensing*, vol. 38, no. 4, pp. 1656–1668, 2000.

[TOR 02] TORRUNGRUENG D., JOHNSON J.T., CHOU H.T., "Some issues related to the novel spectral acceleration method for the fast computation of radiation/scattering from one-dimensional extremely large scale quasi-planar structures", *Radio Science*, vol. 37, no. 2, pp. 1–20, 2002.

[TSA 93a] TSANG L., CHANG C.H., SANGANI H., "A banded matrix iterative approach to Monte Carlo simulations of scattering of waves by large scale random rough surface problems: TM case", *Electronics Letters*, vol. 29, pp. 1666–1667, 1993.

[TSA 93b] TSANG L., CHANG C.H., SANGANI H., *et al.*, "A banded matrix iterative approach to Monte Carlo simulations of large scale random rough surface scattering: TE case", *Journal of Electromagnetic Waves and Applications*, vol. 29, no. 9, pp. 1185–1200, 1993.

[TSA 95] TSANG L., CHANG C.H., PAK K., *et al.*, "Monte-Carlo simulations of large-scale problems of random rough surface scattering and applications to grazing incidence with the BMIA/Canonical grid method", *IEEE Transactions on Antennas and Propagation*, vol. 43, no. 8, pp. 851–859, 1995.

[TSA 00] TSANG L., KONG J.A., DING K.-H., *et al.*, *Scattering of Electromagnetic Waves, Numerical Simulations*, John Wiley & Sons, New York, 2000.

[TSO 10] TSOGKAS G.D., ROUMELIOTIS J.A., SAVAIDIS S.P., "Electromagnetic scattering by an infinite elliptic dielectric cylinder with small eccentricity using perturbative analysis", *IEEE Transactions on Antennas and Propagation*, vol. 58, no. 1, pp. 196–205, 2010.

[VAL 94] VALLE P.J., GONZALEZ F., MORENO F., "Electromagnetic wave scattering from conducting cylindrical structures on flat substrates: study by means of the extinction theorem", *Applied Optics*, vol. 33, pp. 512–523, 1994.

[VID 97] VIDEEN G., NGO D., "Light scattering from a cylinder near a plane interface: theory and comparison with experimental data", *Journal of the Optical Society of America A*, vol. 14, pp. 70–78, 1997.

[WAN 03] WANG X., WANG C.-F., GAN Y.-B.G., *et al.*, "Electromagnetic scattering from a circular target above or below rough surface", *Progress in Electromagnetic Research*, vol. 40, pp. 207–227, 2003.

[WAR 01] WARNICK K.F., CHEW W.C., "Numerical simulation methods for rough surface (topical review)", *Waves in Random Complex Media*, vol. 11, pp. R1–R30, 2001.

[YE 06] YE H., JIN Y.Q., "Fast iterative approach to difference electromagnetic scattering from the target above a rough surface", *IEEE Transactions on Geoscience and Remote Sensing*, vol. 44, no. 1, pp. 108–115, 2006.

[YE 07] YE H., JIN Y.Q., "A hybrid analytic-numerical algorithm of scattering from an object above a rough surface", *IEEE Transactions on Antennas and Propagation*, vol. 45, no. 5, pp. 1174–1180, 2007.

[YOU 88] YOUSSIF H.A., KOHLER S., "Scattering by two penetrable cylinders at oblique incidence. I. The analytical solution", *Journal of the Optical Society of America A*, vol. 5, no. 7, pp. 1085–1096, 1988.

[ZOU 11] ZOUROS G.P., "Electromagnetic plane wave scattering by arbitrarily oriented elliptical dielectric cylinders", *Journal of the Optical Society of America A*, vol. 28, no. 11, pp. 2376–2384, 2011.

[ZOU 13] ZOUROS G.P., "Oblique electromagnetic scattering from lossless or lossy composite elliptical dielectric cylinders", *Journal of the Optical Society of America A*, vol. 30, no. 2, pp. 196–205, 2013.

Index

N, P

R

S

T, W

Printed and bound by CPI Group (UK) Ltd, Croydon, CR0 4YY
13/09/2026
14961081-0007